T0227988

Modern Trends in
# CHEMISTRY AND CHEMICAL ENGINEERING

Modern Trends in
# CHEMISTRY AND CHEMICAL ENGINEERING

*Edited By*
**A. K. Haghi,** PhD
Associate member of University of Ottawa, Canada;
Freelance Science Editor, Montréal, Canada

Apple Academic Press

TORONTO    NEW JERSEY

© 2012 by
Apple Academic Press Inc.
3333 Mistwell Crescent
Oakville, ON L6L 0A2
Canada

Apple Academic Press Inc.
1613 Beaver Dam Road, Suite # 104
Point Pleasant, NJ 08742
USA

First issued in paperback 2021

*Exclusive worldwide distribution by CRC Press, a Taylor & Francis Group*

ISBN 13: 978-1-77463-193-5 (pbk)
ISBN 13: 978-1-926895-00-0 (hbk)

This book contains information obtained from authentic and highly regarded sources. Reprinted material is quoted with permission and sources are indicated. A wide variety of references are listed. Reasonable efforts have been made to publish reliable data and information, but the authors, editors, and the publisher cannot assume responsibility for the validity of all materials or the consequences of their use. The authors, editors, and the publisher have attempted to trace the copyright holders of all material reproduced in this publication and apologize to copyright holders if permission to publish in this form has not been obtained. If any copyright material has not been acknowledged, please write and let us know so we may rectify in any future reprint.

All rights reserved. No part of this work covered by the copyright hereon may be reproduced or used in any form or by any means—graphic, electronic, or mechanical, including photocopying, recording, taping, or information storage and retrieval systems—without the written permission of the publisher.

**Library and Archives Canada Cataloguing in Publication**

Modern trends in chemistry and chemical engineering/[edited by] A.K. Haghi.

Includes bibliographical references and index.
ISBN 978-1-926895-00-0
1. Chemical engineering. 2. Chemistry. I. Haghi, A. K.

TP155.M63 2011          660          C2011-906630-0

**Trademark Notice:** Registered trademark of products or corporate names are used only for explanation and identification without intent to infringe.

Apple Academic Press also publishes its books in a variety of electronic formats. Some content that appears in print may not be available in electronic format. For information about Apple Academic Press products, visit our website at **www.appleacademicpress.com**

# Contents

# List of Contributors

**Minakshi Das**
Department of Basic Sciences and Humanities/Chemistry, Techno Global-Balurgaht, Balurghat-733101.

**Chandra Chur Ghosh**
Department of Basic Science and Humanities/Chemistry and Theoretical and Computational Chemistry Laboratory, Techno Global-Balurghat, Balurghat-733103, India.

**A.K. Haghi**
University of Guilan, Rasht, Iran.

**Nazmul Islam**
Department of Basic Science and Humanities/Chemistry and Theoretical and Computational Chemistry Laboratory, Techno Global-Balurghat, Balurghat-733103, India.

**M. Kanafchian**
University of Guilan, Iran.

**Z. Moridi Mahdieh**
University of Guilan, Rasht, Iran.

**Hamideh Mirbaha**
Department of Textile, University of Guilan, Rasht, Iran.

**Javad Mokhtari**
Department of Textile Engineering, University of Guilan, Rasht-Tehran Road, Rasht, Iran.

**Z. Moridi**
Department of Textile Engineering, Faculty of Engineering, P.O. BOX 3756, University of Guilan, Rasht, Iran.

**V. Mottaghitalab**
Department of Textile Engineering, Faculty of Engineering, P.O. BOX 3756, University of Guilan, Rasht, Iran.

**Mahdi Nouri**
Department of Textile Engineering, University of Guilan, Rasht-Tehran Road, Rasht, Iran.

**N. Piri**
University of Guilan, Rasht, Iran.

**Mohammad Reza Saboktakin**
Baku State University, Azerbaijan.

**Mohammad Seifpoor**
Department of Textile Engineering, University of Guilan, Rasht-Tehran Road, Rasht, Iran.

# List of Abbreviations

| | |
|---|---|
| ABHB | 3, 3′- azobis(6-hydroxy benzoic acid) |
| AChE | Acetylcholinesterase |
| AM | Arithmetic mean |
| AM1 | Austin model 1 |
| 5-ASA | 5-aminosalicylic acid |
| BE | Bond energy |
| BNCs | Bionanocomposites |
| CA | Cross-linking agent |
| CB | Carbon black |
| CDA | Cubane-1, 4-dicarboxylic acid |
| CE | Configuration energy |
| CHT | Chitosan |
| CL | ε-Caprolactone |
| CMC | Carboxymethylcellulose |
| CMS | Carboxymethyl starch |
| CNs | Cellulose nanocrystals |
| CNTs | Carbon nanotubes |
| CRT | Chemical reactivity theory |
| CVD | Chemical vapor deposition |
| DCM | Dichloromethane |
| DD | Degree of deacetylation |
| DFT | Density functional theory |
| DLS | Dynamic light scattering |
| DMA | Dynamic mechanical analysis |
| DMF | Dimethylformamide |
| DS | Degrees of substitutions |
| DSC | Differential scanning calorimetry |
| DWNTs | Double-walled nanotubes |
| EA | Electron affinity |
| FCNs | Flax cellulose nanocrystals |
| FFA | Flufenamic acid |
| FR | Folate receptor |
| FTIR | Fourier transform infrared spectra |
| GAP | Gross atom population |
| GATA | Glucose-6-acrylate-1, 2, 3, 4-tetraacetate |

| | |
|---|---|
| GIT | Gastrointestinal tract |
| GM | Geometric mean |
| GOP | Gross orbital population |
| GSTP | Guilan Science and Technology Park |
| HAP | Hydroxyapatite |
| HEC | Hydroxyethylcellulose |
| HEMA | 2- Hydroxyethyl methacrylate |
| HOMO | Highest occupied molecular orbital |
| HSAB | Hard soft acid base |
| HSE | Heat-separated epidermis |
| IP | Ionization potential |
| LUMO | Lowest unoccupied molecular orbital |
| MAA | Methacrylic acid |
| MMT | Montmorillonite |
| MSA | Maleic starch half-ester acid |
| MWNTs | Multiwalled nanotubes |
| NMR | Nuclear magnetic resonance |
| PAAc | Polyacrylic acid |
| PAN | Polyacrylonitrile |
| PCL | Polycaprolactone |
| PCMs | Phase change materials |
| PEG | Polyethylene glycol |
| PEO | Polyethylene oxide |
| PHA | Poly($\beta$-hydroxyalkanoates) |
| PHO | Poly($\beta$-hydroxyoctanoate) |
| PLA | polylactic acid |
| PMAA-g-St | Polymethacylic acid-graft-starch |
| PPSN | Poly-propylene spun-bond nonwoven |
| PPy | Polypyrrole |
| PS | Plasticized starch |
| PVA | Polyvinyl alcohol |
| PVC | Polyvinyl chloride |
| RBCs | Red blood cells |
| SA | Salicylic acid |
| SBC | Simple bond charge |
| SEM | Scanning electron microscopy |
| SGF | Simulated gastric fluid |
| SIF | Simulated intestinal fluid |
| SPCL | Starch with polycaprolactone |

| | |
|---|---|
| SPI | Silver paint |
| SR | Stability ratio |
| SWNTs | Single walled nanotubes |
| TDDS | Transdermal drug delivery systems |
| TFA | Triflouroacetic acid |
| TPP | Tetraphenylporphyrin |
| TPS | Thermoplastic starch |
| TSDC | Thermally stimulated depolarization current |
| TU | Thermochemical unit |
| w/w | Water-in-water |
| WAXD | Wide-angle X-ray diffraction |
| ZDO | Zero differential overlap |

# Preface

This new book presents and discusses current research done in the field of chemistry. In Chapter 1, time evolution of the electronegativity is discussed. Various scales of electronegativity like Pauling's Quantum thermo-chemical scale of electronegativity, Malone's Scale of electronegativity, Walsh's Scale of electronegativity, and so forth are mentioned. Authors also provide inter-relationship between the electronegativity and other periodic parameters. In Chapter 2 time evolution of the electronegativity is discussed. Electronegativity equalization principle becomes one of the most useful applications of the electronegativity. It includes dipole charge and dipole moment in terms of electronegativity, electronegativity and the HSAB principle, and so forth. Chapter 3 focuses on the starch nanocomposite and nanoparticles and its biomedical applications. The author further discusses about the modification of starch. Chapter 4 has described the updates on lamination of nanofiber. Authors prepared a surface image of nanofiber web after laminating at different temperature using an optical microscope. It was observed that nanofiber web was approximately unchanged when laminating temperature was below Poly-propylene Spun-bond Nonwoven (PPSN) melting point. Chapter 5 includes electrospinning of chitosan (CHT). The mechanical and electrical properties of neat CHT electrospun natural nanofiber mat can be improved by addition of the synthetic materials including carbon nanotubes (CNTs). Dynamic light scattering (DLS) is a sophisticated technique used for evaluation of particle size distribution. In Chapter 6, smart nanofibers based on nylon 6,6/polyethylene glycol blend are discussed. Thermal properties of electrospun nanofibers examined with differential scanning calorimetry (DSC). It is clear that increasing the polyethylene glycol (PEG) content in the blend nanofibers has a little effect on the phase change temperatures, but strongly affects the latent heat of phase changes. In Chapter 7, authors have explained nano-biocomposites with chitosan matrix. They also explained carbon nanotubes (CNTs) which are straight segments of tube with arrangements of carbon hexagonal units. CNTs can be classified as single walled carbon nanotubes (SWNTs) and multi walled carbon nanotubes (MWNTs). Chapter 8 discusses about polypyrrole coated polyacrilonitril electrospun nanofibers. Authors' observed problems on application of conducting polymers have been brittleness, insolubility, and unstable electrical properties. Fiber formation and morphology of the coated nanofibers were determined using a scanning electron microscope (SEM). Chapter 9 focuses on semi-empirical AM-1 studies on porphyrin which include global reactivity parameters, local reactivity parameters, and atomic charge. Authors have calculated the eigen values and eigen functions of molecules in the chapter.

— **A. K. Haghi**

# Chapter 1

## Time Evolution of the Electronegativity Part-1: Concepts and Scales

Nazmul Islam and Chandra Chur Ghosh

---

### INTRODUCTION

The concept of electronegativity had been a part of chemical thought for nearly about 140 years. It is opined [1] that no concept more thoroughly encompasses the fabric of modern chemistry that that of electronegativity. Nowaday, it is established that the electronegativity is an indispensable tool in every branch (both theoretical and experimental) of chemistry, physics, engineering, and biology.

The concept of electronegativity was instigated in 1809 when Avogadro [2–4] pointed out the similarities between the acid-base neutralization process and the electrical charge neutralization process. Avogadro proposed an "oxygenicity scale" on which elements were placed depending upon their tendency to react with other elements. Thereafter, Berzelius [5–9] first coined the term "electronegativity" instead of "oxygenicity" and formulated a "universal scale of electronegativity" of the elements. Berzelius [5, 6] further categorized elements into two classes: (a) electronegative and (b) electropositive. Later it was established that the electronegativity data of elements computed using Berzelius' Scale correlate remarkably well with the electronegativity data computed using the scale of Pauling [10, 11] which was based on thermochemical data and also the scale of Allred and Rochow [12] which was based on the force concept. Thus, the term electronegativity and its association with an electron attracting power between atoms originated with J. J. Berzelius in 1811, and its continuous use since suggests that a true chemical entity is manifest itself.

However, Berzelius' theory failed to account for half of all possible chemical reactions such as endothermic associations and exothermic dissociations. Moreover, Berzelius' theory could not account for increasingly complex organic molecules, and also it is incompatible with Faraday's laws of electrolysis [1].

Pauling [10, 11] first gave the objection for the use of electrode potential as a measure of electron attracting power. Then, based on thermochemical data and quantum mechanical arguments, Pauling [10, 11] defined electronegativity as "the power of an atom in a molecule to attract electron pair toward itself." Electronegativity is a fundamental descriptor of atoms molecules and ions which can be used in correlating a vast field of chemical knowledge and experience. Allen [13, 14] considered electronegativity as the configuration energy of the system and argued that electronegativity is a fundamental atomic property and is the missing third dimension to the periodic table. He further assigned electronegativity as an "ad hoc" parameter. Huheey, Keiter, and Keiter [15] opined that the concept of electronegativity is simultaneously one of the most important and difficult problems in chemistry. Frenking and Krapp [16] opined that the appearance and the significance of the concepts like the electronegativity resembles the "unicorns of mythical saga," which has no physical sense but without the concept and operational significance of which chemistry becomes disordered and the long established unique order in chemico-physical world will be taken aback [17–22]. Fukui [23] opined that the static and dynamic behavior of molecules can be well understood by the use of the electronegativity concept. The fundamental quantities of inorganic, organic, and physical chemistry such as bond energy, polarity, and the inductive effect can be visualized in terms of electronegativity. At present, the concept of electronegativity is not only widely used in chemistry but also in biology, physics, and geology [24–26]. An outstanding dependence of the superconducting transition temperature on electronegativity is found for both solid elements and high-temperature superconductors [27–29]. Electronegativity concept has also been successfully used to correlate various spectroscopic phenomenons such as nuclear quadruple coupling from microwave and radio wave frequency spectroscopy [30] and with the chemical shift in nuclear magnetic resonance spectroscopy [31] and so forth. Lackner and Zweig [32] pointed out that the electronegativity has led to the correlation of vast number of important atomic and molecular properties and also to the qualitative understanding of quark atoms. The concept of electronegativity has been successively used by scientists to explain the geometry and properties of molecule such as superconductivity, photocatalytic activity, magnetic property, and optical basicity [33–37]. Furthermore, in recent years, electronegativity concept has been used to design materials [38] and drugs [39].

The intent of this work is to try to recapitulate the time evolution of the scales and concepts of electronegativity.

## VARIOUS SCALES OF ELECTRONEGATIVITY

Innumerable works of chemists from abundance of chemical observations has filled up the field of electronegativity. Chemists have been able to derive ingenious concepts and scales of electronegativity that have proved their usefulness in predicting and systematizing chemical facts. In principle, pure chemical knowledge and experience allows a reasonable estimation of electronegativity character of atoms, yet translation of that knowledge into some numerical indexing has been the target of innumerable workers. As a result of these intellectual exercises, ever since the concept of electronegativity was presented by Pauling, the useful hypothetical or qualitative entities like the electronegativity which were abstract semiotic representations can be considered as theoretical quantities of cognitive representations. However, scientific world till now, believe that the final scale of electronegativity is not proposed by any one. Electronegativity is empirical and will empirical as there is no quantum mechanical operator for it and also electronegativity is not an experimentally measurable quantity [17–20, 40, 41]. In this section, we reviewed some of the most important and useful scales of electronegativity of atoms, ions, and orbitals.

## Pauling's Quantum Thermo-chemical Scale of Electronegativity [10]

Pauling [10, 11] by an ingenious mixing of thermodynamical and quantum mechanical arguments proposed the word "electronegativity" as *"the power of an atom in a molecule to attract electrons toward itself."* During research on hetero nuclear diatomics, Pauling discovered that the properties related to the energy and charge distribution in chemical bonds between hetero atoms can be correlated with some internal constituent of atoms which forms the hetero nuclear bonds. The properties include ionic character, the charge distribution, the degree of polarity, the bond dissociation energies, bond moments, force constants, and the like. Thus, the treatment of heteronuclear bonds revolves around the concept of electronegativity and the use of electronegativity to understand bond energy differences was widely appreciated. Pauling supposed that the energy of an ordinary covalent bond X-Y is generally larger than the additive mean of the energies of the bond X-X and Y-Y and the enhancement factor $\Delta$, increases as the atoms X and Y become more and more unlike in their electronegativity property. Considering the electronegativities of X and Y are $\chi_X$ and $\chi_Y$, Pauling [10, 11] proposed the relationship between the electronegativity difference and the enhancement factor as

$$\chi_X \sim \chi_Y = 0.208\sqrt{\Delta} \tag{1}$$

The enhancement factor $\Delta$, calculated by Pauling as

$$\Delta = D_{(X-Y)} - 0.5[D_{(X-X)} + D_{(Y-Y)}] \tag{2}$$

where the dissociation energies, D's, of the X-Y, X-X, and Y-Y bonds are expressed in eV unit.

The unit of electronegativity in Pauling Scale is (energy)$^{1/2}$. Now this unit is referred as thermochemical unit (TU).

Pauling [10, 11] computed electronegativity values for 33 elements. Thereafter, a number of workers revisited and extended the Pauling's Scale. For example, Haissinsky [42] extended Pauling's calculations to 73 elements. Haissinsky [42] also showed that for multivalent elements, electronegativity is a function of valency of the atoms. Huggins [43] re-evaluated the electronegativities of 17 elements of the periodic table. Gordy and Orville Thomas [44] pointed out that the Huggins' electronegativity values [43] are generally higher than Pauling's electronegativity values. They demonstrated that if the Huggins' electronegativity values are downgraded by the factor 0.1 and the resulting values are round off to two significant figures then Huggins' electronegativity values agree well with the Pauling's values.

Altshuller [45] evaluated electronegativity data of the Copper, Zinc, and Gallium sub group elements. Thereafter, Allred [46] revisited the Pauling's Electronegativity Scale and calculated the electronegativity data of 69 elements using corresponding thermochemical data of the elements published at that time. Altshuller [45] also summarized the trends of electronegativity values within the periodic system. A theoretical basis of Pauling's Scale was given by Mulliken [47].

It is apparent from Pauling's definition that electronegativity is not the property of isolated atom, but it depends on the molecular environment in which the atom is present, that is, electronegativity is a property of atoms arises when the atoms form molecules. But latter, it is established that electronegativity is an intrinsic property of a free atom [13, 14, 22, 25, 48–53].

## Malone's Scale of Electronegativity [54]

Just 1 year after the announcement of the electronegativity concept by Pauling [10], Malone [54] suggested a relationship between the dipole moment in Debye ($\mu_d$) of a hetaronuclear bond X-Y and the electronegativity difference, $\chi_X \sim \chi_Y$, as:

$$\chi_X \sim \chi_Y \, \mu \, \mu_d \tag{3}$$

A deeper study on the Malone's Scale reveals that the scale can be applied remarkably well in a few well known cases (e.g., hydrogen halides) but in case of a majority of compounds due to the inaccuracy in the computed result this scale cannot be accepted as a reasonable scale of electronegativity.

## Mulliken's Scale of Electronegativity [55]

In 1934, an empirical spectroscopic definition of electronegativity was proposed by Mulliken [55] as the average of the *IP* and *EA* for the valence state of an atom.

Mulliken considered two limiting resonance structures of the diatomic complex XY.

$$X^{\partial+} Y^{\partial-} \leftrightarrow XY \leftrightarrow X^{\partial-} Y^{\partial+} \tag{4}$$

If one replaces the limiting structures by the equivalent ionic components then equation (4) looks like:

$$X^+ + Y^- \leftrightarrow X + Y \leftrightarrow X^- + Y^+ \tag{5}$$

Case-1: Y is more electronegative than X, Y holds more negative charge than X that is:

$$X + Y \rightarrow X^+ + Y^- \tag{6}$$

Energy change associated with the reaction (6) is given by the difference between the energy required to remove an electron from A, its ionization potential (*IP*), and the energy consumed to attach the electron to the outer shell of B, its electron affinity (*EA*)

$$\Delta E_{(X^+ Y^-)} = IP_X - EA_Y \tag{7}$$

Case-2: X is more electronegative than Y, X holds more negative charge than Y that is,

$$X + Y \rightarrow X^- + Y^+ \tag{8}$$

The consumed energy is

$$\Delta E_{(X^- Y^+)} = IP_Y - EA_X \tag{9}$$

Now, Mulliken's assumption was that the difference between $A^+B^-$ and $A^-B^+$ can be neglected as they are not truly ionic. So the involved energies, $\Delta E_{(A^+B^-)}$, $\Delta E_{(A^-B^+)}$ can be equalized.

$$\Delta E_{(X^+Y^-)} = \Delta E_{(X^-Y^+)} \tag{10}$$

that is, $IP_X - EA_Y = IP_Y - EA_X$ $\qquad(11)$

or, $IP_X + EA_X = IP_Y + EA_Y$ $\qquad(12)$

The equation (12) reveals that the sum of ionization energy and electron affinity of each separate atom becomes equal when they are combined to form the complex, XY.

Hund [56] stated that the quantities average of $IP$ and $EA$, that is, $(IP + EA)/2$, is an approximate criterion for their equal participation in a chemical bond.

Using this idea, Mulliken [55] took an arithmetic mean of the first ionization energy and electron affinity as a qualitative definition of electronegativity for any species X (atom, molecule, or radical in its state of interaction):

$$\chi_X \approx (IP_X + EA_X)/2 \tag{13}$$

It is more usual to use a linear transformation to transform these absolute values into values which resemble the more familiar Pauling values. Plotting the $(I + A)$ with Pauling electronegativity values, the electronegativity scale was designed as

$$\chi = a\,(IP + EA) + b \tag{14}$$

where a and b are the constants.

Putting "$IP$" and "$EA$" in electron volt and using the method of least square fitting, the "a" and "b" values are computed as $a = 0.187$ and $b = 0.17$.

Coulson [57] opined that Mulliken's measure of electronegativity is better and more precise than Pauling's electronegativity data.

Mulliken's Electronegativity Scale is absolute and more fundamental because it only depends on the fundamental energy value of the isolated atom. Also, it is more precise because it bears the modern density functional definition of electronegativity [58, 59].

$$\chi_{DFT} = -(\partial E/\partial N)_v \tag{15}$$

From the energy versus number of electron curve (E vs. N curve), it is transparent that the change in energy, $\Delta E$, is associated with two electrons changes. If we consider S as a neutral species having energy $E_N$, and having a total number of N electrons, then the corresponding cation and anion, $S^+$ and $S^-$ have the energy $E_{N-1}$ and $E_{N+1}$ and the number of electrons $N-1$ and $N+1$ respectively.

Putz [48] showed that the density function electronegativity ($\chi_{DFT}$) approximates the former Mulliken electronegativity formula ($\chi_M$).

$$\chi_{DFT} = -(\partial E/\partial N)_v = -(E_{N+1} - E_{N-1})/2 = (IP + EA)/2 = \chi_M \tag{16}$$

Bratsch [60, 61] revisited the theoretical basis, concept and application of Mulliken electronegativity in terms of valence state promotional energies. He considered the valence state ionizational potential $(IP_v)$ and electron affinity $(EA_v)$ proposed by Hinze and Jaffe [62, 63] as:

$$IP_v = IP + P^+ - P^0 \tag{17}$$

and,

$$EA_v = EA + P^0 - P^- \tag{18}$$

where P stands for valence shell promotional energy.

Bratsch [60, 61] showed that the Mulliken "a" and "b" parameters for a given element vary linearly with the increasing degree of "s" character. Bratsch [60, 61] further opined that a linear relationship between Mulliken and Pauling electronegativity is not possible to propose because of the dimension mismatch in the two scales. Mulliken's electronegativity has the dimension of energy while the Pauling Scale has the dimension of the square root of energy. Bratsch [60, 61] corrected this dimensional mismatch and proposed a linear relationship between the Pauling's electronegativity$(\chi_P)$ and square root of Mulliken's electronegativity$(\chi_M)$ as follows:

$$\chi_P = 1.35(\chi_M)^{1/2} - 1.37 \tag{19}$$

Using the computed electronegativity data, Bratsch [60, 61] computed the partial ionic charge, the bond energy and the group electronegativity for some systems and also connected the correlation coefficients "a" and "b", with the essence of the Hard Soft Acid Base (HSAB) principle of Pearson [64].

### Gordy's Scale of Electronegativity [65]

Gordy [65] suggested that the electronegativity of an atom $(\chi_G)$ is the electrostatic potential (or the effective nuclear charge $Z_{eff}$, of the nucleus on the outermost electron) felt by one of its valence electrons at a radial distance equal to atom's single bond covalent radius$(r_{cov})$.

that is,

$$\chi_G = e\, Z_{eff}/r_{cov} \tag{20}$$

The electrostatic electronegativity scale of Gordy [65] was scientifically justified by a good number of workers viz. Pasternak [66], Ray, Samuels, and Parr [67], Politzer, Parr, and Murphy [68]. Gordy and Orville Thomas [44] pointed out that the electronegativity ansatz of Gordy cannot be used to calculate the electronegativity data of the transition elements for which the energy levels of different shells begin to overlap. To explain the deviation Gordy and Orville Thomas [44] modified the scale proposed by Gordy. Gordy and Orville Thomas [44] postulated that the effective nuclear charge $Z_{eff}$, can be obtained with the approximation that all electrons are packed in closed shells below the valence shells and these packed electrons use their full screening power to all the valence electrons which exert equal screening.

Gordy and Orville Thomas [44] proposed the following expression to compute the effective nuclear charge $(Z_{eff})_{GT}$, as

$$(Z_{eff})_{GT} \approx n - \sigma(n - 1) \tag{21}$$

where n is the number of electron in the valence shell of the atom, $\sigma$ is the screening constant of the valence electrons.

Substituting the $Z_{eff}$ by $(Z_{eff})_{GT}$ in equation (20) the electronegativity ansatz of Gordy is rewritten by Gordy and Orville Thomas [44] as:

$$\chi_{GT} = e\{n - \sigma (n - 1)\}/r_{cov} \tag{22}$$

Ghosh and Chakraborty [53] pointed out that $r_{cov}$ cannot be used as a necessary input in computing electronegativity as electrostatic potential. Ghosh and Chakraborty [53] modified Gordy's formula by substituting covalent radii by absolute radii. They also proposed that the electronegativity, $\chi_{GC}$, is not equal, rather proportional to $Z_{eff}/r$. Thus, the modified electronegativity ansatz is:

$$\chi_{GC} = a(Z_{eff}/r_{abs}) + b \tag{23}$$

where "a" and "b" are the constants for a given period.

Recently, Islam [69] showed that the Gordy's Scale of atomic electronegativity can be derived relying upon the charge sphere model for IP and EA. This study further reveals that the three definitions of electronegativity—the density functional definition ($\chi_{DFT}$), the Mulliken's definition ($\chi_M$) and the Gordy's definition ($\chi_G$) are nicely converged to a single point.

$$\chi_{DFT} = -(\partial E/\partial N)_v = -(E_{N+1} - E_{N-1})/2 = (IP + EA)/2 = \chi_M = (Z_{eff}/r_{abs}) = \chi_G \tag{24}$$

### Walsh's Scale of Electronegativity [70]

In 1952, Walsh [70] correlated electronegativity and stretching force constants of the bonds between an atom and a hydrogen atom. Walsh [70] proposed that the electronegativity of an atom or any group "X" is the stretching force constants of its bonds to a hydrogen atom (X-H) and also demonstrated very clearly that polarity does not increase bond strength, a conclusion which might have been drawn from the original arguments of Pauling [10].

### Sanderson's Scale of Electronegativity (1952)

Sanderson [71] noted the inter-relationship between the electronegativity and the atomic size, and has proposed a method of evaluation of electronegativity based on the reciprocal of the atomic volume. With knowledge of bond lengths, Sanderson's method allows to estimate the bond energies in a wide range of compounds.

Focus on the chemical bond that hold together the atoms that form the molecules, an answer of the fundamental question, why do atoms interact to form molecule was given by the electronegativity equalization principle. After the announcement of the very fundamental law of nature—the electronegativity equalization principle by Sanderson [71], it becomes one of the most useful applications of the electronegativity. To formulate the electronegativity equalization principle Sanderson [71] stated that when two atoms having different electronegativity come together to form a molecule, the electronegativities of the constituent atoms become equal, yielding the molecular equalized electronegativity. Thus for the first time the concept of electronegativity had been thought of as a dynamic property rather than a static one. Electrons in a stable homonuclear covalent bond are equally attracted to both nuclei. But this is not true in case of a heteronuclear system, where two atoms (or more) having different electro- negativity values are joined through covalent bond. The more electronegative atom having more electron attracting power attracts the bonding electron pair more towards itself. Thus some amount of charge transferred from the lower electronegative atom to the higher one. This can be also viewed as charge is transferred from the atom having higher chemical potential value to the atom having lower chemical potential value un- til both the chemical potential and electronegativity of the constituents becomes equal.

Two different electronegative atoms have atomic orbitals of different energies. The process of bond formation must provide a pathway by which the energies of the bond- ing orbitals become equalized. If in the bond formation process the electronegativity of the higher electronegative atom decrease as that atom acquires electronic charge ($\delta$) and that of lower electronegative atom increase as it loses the electronic charge ($\delta$). Sanderson [71] postulated a geometric mean principle for the electronegativity equal- ization. He pointed out that the final electronegativity of a molecule is the geometric mean of the original atomic electronegativities. The electronegativity equalization principle is now linked to the fundamental quantum mechanical variation principle. Parr et al. [58] identified electronegativity as the amount of energy required to remove a small amount of electron density from the molecule at the point r, that is,

$$\chi(r) = \delta Ev(\rho)/\delta\rho(r) \qquad (25)$$

Parr et al. [58] have noted that the energy is minimized only if the electronegativ- ity is equalized, because if there are two place in the molecule with different electro- negativity, then transferring a small amount of electron density, q, from the place to lower electronegativity ($r_<$) to the place with greater electronegativity ($r_>$) will lower the energy. Parr and Bartolotti [72] gave a proof of the electronegativity equalization principle from a sound density functional theoretical [73, 74], background. The term chemical potential as it occurs in thermodynamics [75] has long been accepted as a perspicuous description of the escaping tendency of a component from a phase. Parr et al. [58] identified electronegativity as the negative of the chemical potential of the system. They also pointed out that both parameters can be adopted at the molecular level because they have the very same properties in the charge equalization procedure. Thus they suggested that both the words, "electronegativity" and "chemical potential," can be applied for the electronegativity equalization procedure but they prefer the lat- ter for their discussion.

They correlated Charge Transfer, Electronegativity Difference, and Energy Effect of Charge Transfer with the geometric mean principle of electronegativity equaliza- tion [71].

One can use this equalization concept as a guide to the outcome of metathesis reac- tions. This principle can be used to calculate various physic-chemical properties of the atoms in the molecule and molecular properties like the partial charge of the atoms and groups, dipole moment, bond distance, and so forth.

### Allred and Rochow's Scale of Electronegativity [12]

Allred and Rochow [12] postulated that electronegativity of an atom is proportional to the charge experienced by an electron on the outermost shell of an atom. The higher the charge of atomic surfaces per unit area, the greater the tendency of that atom to attract electrons.

Now, the charge experienced by an electron on the surface of an atom or on the outermost shell of an atom can be described in terms of the effective nuclear charge, $Z_{eff}$ experienced by valence electrons and the surface area of the atom. Now, as the surface area of an atom is proportional to the covalent radius of the atom, the electronegativity can be represented as

$$\chi \mu \; Z_{eff}/r^2_{cov} \tag{26}$$

Allred and Rochow [12] suggested a linear relationship between $\chi$ and $Z_{eff}/r^2_{cov}$ as,

$$\chi = a(Z_{eff}/r^2_{cov}) + b \tag{27}$$

where "a" and "b" are the correlation constants.

Scaling with Pauling electronegativity values [10, 11], Allred and Rochow [12] proposed a new electronegativity($\chi_{AR}$) scale as:

$$\chi_{AR} = 0.359(Z_{eff}/r^2_{cov}) + 0.744 \tag{28}$$

The concept of Allred and Rochow [12] was justified by a good number of workers. For example, Little and Jones [76] verified and recalculated the electronegativity of the atoms of the periodic table based on the force concept of Allred and Rochow [12].

Mande, Deshmukh, and Deshmukh [77] on the basis of relativistic Dirac equation, calculated screening constants using X-ray spectroscopic method and using the spectroscopic effective nuclear charge of the atoms for the valence states they evaluated the electronegativity of the atoms by the following modified ansatz:

$$\chi = 0.778 \; (Z_{eff})_{spectroscopic}/r^2_{cov}) + 0.5 \tag{29}$$

The constants of the above ansatz (29) was evaluated by Mande et al. [77] by plotting $(Z_{eff})_{spectroscopic}/r^2_{cov}$ with the Pauling's electronegativity values [10, 11].

The electrostatic scale of Allred and Rochow [12] was further modified by Boyd and Markus [78]. Boyd and Markus [78] proposed a non-empirical electrostatic model for calculating the attraction between the screened nucleus and an electron at a distance corresponding to the relative radius of the atom, that is, the electronegativity $\chi_{BM,}$ as:

$$\chi_{BM} = \frac{kz}{r^2}[1 - \int_0^r D(r)\,dr] \tag{30}$$

where Z is the atomic number, r is the relative radius of the atom, D(r) is the radial density function and k is a constant so chosen(69.4793au) that the electronegativity value of F becomes 4. Boyd and Markus [78] calculated the relative radius of atom, r using the density contour approach of Boyd [79] on the basis of analytical Hartree–Fock wave function of atoms proposed by Clementi and Roetti [80].

In a recent work, Ghosh et al. [52] have pointed out some inconsistency in the Allred and Rochow's Electronegativity Scale [12] and also in the previous modifications of Allred and Rochow's Scale [12].

They found that:

1. Although, Allred and Rochow [12], Mande et al. [77], and Little and Jones [76] used the force concept to evaluate the electronegativity of atoms but the dimension of the electronegativity is not be mentioned by any of them.

2. Allred and Rochow [12], Mande et al. [77], Little and Jones [76] used the covalent radii in atomic unit to calculate the electronegativity values in their proposed electronegativity scale.

However, these considerations do not compute the electronegativity in force unit.

3. The absolute radius is the true descriptor of atomic electronegativity not the covalent radius.

Ghosh et al. [52] replaced the covalent radii by absolute radii and solved the dimension problem of the Allred and Rochow Scale by proper dimension matching and they reported electronegativity as force.

$$\chi = \text{Force} = e^2 \, (Z_{eff})/r_{abs}^2 \tag{31}$$

## Iczkowski and Margrave's Scale of Electronegativity [81]

Iczkowski and Margrave [81] considered electronegativity as a property of an isolated gaseous atom or ion. By plotting (Fig. 1.1) the atomic energy change with degree of ionization of a chemical system, Iczkowski and Margrave [81] discovered the energy expression for the of an atom, X as:

$$E(N)_X = aN + bN^2 + cN^3 + dN^4 + \dots \tag{32}$$

where N is the number of electrons in the valence shell of nucleus X, and a, b, c and d are the coefficients.

They identified the electronegativity, for an atom or ion, as the slope at the origin - (dE/dN), of the E versus N curve.

$$\chi = -(dE/dN) \tag{33}$$

Klopman [82] postulated that "the atomic terms for any atom can be defined as the sum of those integrals in which the Hamiltonian represents the interaction of the core of the atom with the electron around it" and critically justified the Iczkowski and Margrave's electronegativity concept [81] as under:

The E versus N relationship is usually linear so the higher terms of the equation (32) can be neglected. This leads to the simplified expression (34)

$$E = aN + bN^2 \tag{34}$$

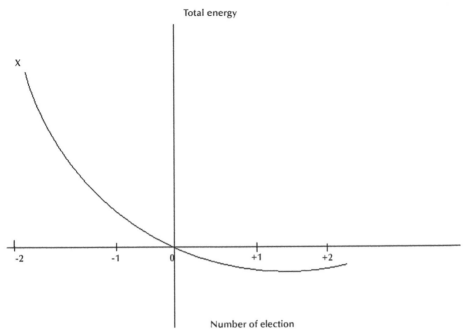

**Figure 1.1.** Plot of the energy change with the degree of ionization.

On differentiation with respect to N, we get:

$$\partial E/\partial N = a + 2b\,N \tag{35}$$

The electronegativity according to the Pauling definition can be represented [82] by the potential around the atom thus it can be represented by $(\partial E/\partial N)$.

$$\chi = (\partial E/\partial N) = a + 2b\,N \tag{36}$$

Iczkowski and Margrave's definition of electronegativity [81] is widely accepted. Hinze, Whitehead, and Jaffe [63] opined that electronegativity is not an atomic property, but the property of an orbital of an atom (X or Y) in a molecule (XY) and thus it is dependent on the valence state of the atom (X or Y). Klopman [82] also opined that electronegativity is an orbital characteristic and therefore the both the ionization potential and electron affinity have to be measured for the same orbital.

Now, for the valence state of an atom, A, N = 1,

$$E = a + b = IP \tag{37}$$

where $a = (3IP - EA)/2$ and $b = (EA - IP)/2$
and for the valence state of an anion, $A^-$, N = 2,

$$E = 2a + 4b = IP + EA \tag{38}$$

Here "*IP*" is the ionization energy and "*EA*" is the electron affinity of the atom (A or B) in its valence state.

When N = 1, the electronegativity leads to an expression similar to that proposed by Mulliken [55].

$$\chi = (\partial E/\partial N)_{N=1} = a + 2b = (IP + EA)/2 \tag{39}$$

Finally Klopman [82] pointed out that however, that in order to represent the electronegativity of an atom in a molecule correctly, the atom must be considered in its valency state and this requires the introduction of electron spin.

## Hinze and Jaffe's Scale for Orbital Electronegativity of Neutral Atoms [62]

Hinze and Jaffe [62] opined that electronegativity is not a property of atoms in their ground state, but of atoms in the same condition in which they are found in molecules, that is, in their valance state. They also noticed that the electronegativity can be defined in terms of bonding orbital and the term "Orbital electronegativity" is then suggested.

In the next year Hinze et al. [63] proposed that "*the power of an atom to attract electrons in a given orbital to itself*" can be correlated with the orbital electronegativity. The orbital electronegativity is then defined as the derivative of the energy of the atom respect to the charge in the orbital, that is, the number of electrons in the orbitals:

$$\chi_j = \partial E/\partial n_j \tag{39}$$

where $\chi_j$ is the orbital electronegativity of the jth orbital and $n_j$ is the occupation number of the jth orbital.

This definition implies two assumptions—(a) that the occupation number $n_j$ may have both integral and non-integral values and, (b) that once assumption (a) is made, than the energy E is a continuous and differentiable function of $n_j$.
Thus,

$$\chi_j = \partial E/\partial n_j = b + 2c\, n_j \tag{40}$$

where b and c are the constants.

## Yuan's Scale of Electronegativity [83]

Yuan [83] defined electronegativity as the ratio of the number of valence electron to the covalent radius. This scale was later modified by Luo and Benson [84–86] on the basis of the number of valance electrons in the bonding atoms and covalent radius of the atom.

## Gyftopoulos and Hatsopoulos's Quantum Thermodynamic Scale of Electronegativity [75]

Gyftopoulos and Hatsopoulos [75] identified electronegativity as the additive inverse of the chemical potential, $\mu$.

Gyftopoulos and Hatsopoulos [75] defined the electrochemical potential of a thermodynamic system as

$$\mu = (\partial E/\partial N)_{entropy} \tag{41}$$

where N is the number of electron

and

$$\chi = -\mu = -(\partial E/\partial N)_{entropy} \qquad (42)$$

### St. John and Bloch's Quantum Defect Scale of Electronegativity [87]

St. John and Bloch [87] suggested that some dimensionless parameters, which are directly derived from atomic spectral data, can be used to define a scale of electronegativity for non-transition elements. St. John and Bloch [87] demonstrated that the Pauling force potential model, which provides the solution of one electron Schrodinger equation, can be successfully applied for the studies of the physico-chemical behavior of the atoms and molecules. The so called "quantum defect" which is physically related to the depth of the potential well and the strength of the effective centrifugal barrier is automatically subsumed in the the eigen values obtained by solving one electron Schrodinger wave equation. Relying on the Pauling's potential force model, St. John and Bloch [87] further opined that it is particularly convenient to express some dimensionless parameters in terms of the positions of the radial maxima of the unscreened, lowest valence eigen functions. For the non-transition elements the s-p hybridization can be reflected in a structural index, S [88]. Now St. John and Bloch [87] redefined the term "S" in terms of the orbital electronegativity as:

$$S = (\chi_0 - \chi_1)/\chi_0 \qquad (43)$$

where $\chi l$ is the orbital components which measures the scattering power of the core for the lth particle wave.

The sum of the orbital components of the electronegativity is proportional to the total electronegativity of the atom.

$$x^{\alpha} \sum_{l=0}^{2} x_l \qquad (44)$$

St. John and Bloch [87] found a linear relationship between the sum of the orbital components of the electronegativity and the Pauling electronegativity data and proposed the quantum defect scale of electronegativity as:

$$x = 0.43 \sum_{l=0}^{2} x_l + 0.24 \qquad (45)$$

St. John and Bloch [87] applied their scale for explaining the structures, chemical properties, and so forth for the simple solids such as zinc-blende and wurtzite.

### Density Functional Scales of Electronegativity (1978)

*Parr et al.'s Scale of electronegativity [58]*

Electronegativity has been one of the most accepted and used concepts in chemistry for more than 60 years however, its physical significance has been elucidated in terms of the density functional theory [73, 74] by Parr et al. [58], who, following Iczkowski and Margrave [81], have demonstrated the electronegativity as the negative of the chemical potential of any system-atom, ion, and molecule.

$$\chi = -\mu = (\partial E / \partial N)_v \qquad (46)$$

where $\mu$ is the chemical potential of the system.

Parr et al. [58] also demonstrated that electronegativity is constant throughout an atom or a molecule. This invention justified and validated Sanderson's electronegativity equalization principle: "when two or more atoms having different electronegativity combine to form a molecule, their electronegativities get equalized" [71].

*Parr and Pearson's Scale of electronegativity based on finite difference approximation [59]*

Parr and Pearson [59], using the method of finite difference approximation, made an attempt to propose an analytical form of electronegativity and hardness on the basis of DFT. The concept of chemical hardness is very old in chemistry whose basis lies on some experimental observations by various inorganic chemists. Hardness (or inverse of hardness, known as "softness") is an intrinsic property of atoms and molecules which signify the deformability of atoms and molecules under small perturbation. More discussion on the hardness is out of scope of this work. However, we proceed to revisit the inter-relationship between DFT, electronegativity, and hardness simultaneously in this section.

The chemical hardness, electronegativity, and DFT came together in the year of 1983 at the Institute for Theoretical Physics in Santa Barbara. It was a great step forward when Parr showed Fig. 1.2 to Pearson where the total electronic energy of a chemical system in its different states of oxidation that is, positive, neutral, and negative states is plotted as a function of number of electrons of those systems. Parr asked him whether the curvature of the curve, in the way in which the slope changes with N, that is, $(\delta^2 E / \delta N^2)_V$ is related to hardness.

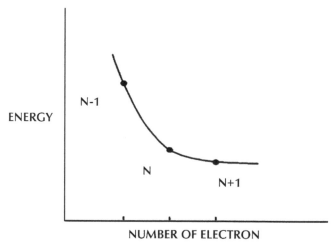

**Figure 1.2.** Plot of total electronic energy (all are negative) of a system in positive (+1), neutral (0), and negative (–1) state as a function of number of electrons (N).

Pearson applied the finite difference approximation method and found an operational formula for $(\delta^2 E/\delta N^2)_v$ which was $(IP–EA)$. Pearson exclaimed that it was exactly what he meant by hardness in his landmark "Hard Soft Acid Base" paper [64]! Then Parr and Pearson, using the density functional theory (DFT) as basis, have rigorously defined the term hardness as,

$$\eta = \tfrac{1}{2}(\delta^2 E/\delta N^2)_v \qquad (47)$$

The softness(S) is defined [89] as the reciprocal of the hardness:

$$S = (1/\eta) \qquad (48)$$

Quantum mechanics provides us to write the energy of the valence electrons in the form of the quadratic approximate equation (34), $E = aN + b\,N^2$, where "a" is a constant-a combination of core integral and a valence shell electron pair repulsion integral and "b" is the half of the average valence shell electron–electron repulsion integral.

Now differentiating equation (34) with respect to N at constant external potential v, Parr and Pearson [59] proposed:

$$(\partial E/\partial N)v = -a - 2bN = (IP + EA)/2 = \chi_M \qquad (49)$$

This electronegativity scale is known as absolute scale of electronegativity.

*Pearson Frontier Orbital scale of lectronegativity [90]*
In 1986, within the limitations of Koopmans' theorem, Pearson [90] putted electronegativity into a MO framework.

The orbital energies of the Frontier Orbitals are given by

$$\varepsilon_{HOMO} = IP \qquad (50)$$

and

$$\varepsilon_{LUMO} = EA \qquad (51)$$

Thus on the basis of frontal orbital theory, he achieved

$$\chi = (\partial E/\partial N)_v = (IP + EA)/2 = -(\varepsilon_{LUMO} + \varepsilon_{HOMO})/2 \qquad (52)$$

## Pasternak's Scale of Electronegativity (1978)

Pasternak [66] using the simple bond charge (SBC) model of diatomic molecule proposed that electronegativity of the atom X and Y in the XY molecule is proportional to the ratio of the nuclear charge of the atom X and Y and half of the bond lengths in XX and YY respectively.

$$\chi_X = C(Z_X/r_X) \qquad (53)$$

and,

$$\chi_Y = C(Z_Y/r_Y) \qquad (54)$$

where C is a constant depends on bond type.

### Zhang's Scale of Electronegativity [91, 92]

Zhang's [91, 92] Scale is based on the ionization energies and the covalent radius of the atom. Zhang defined the term "electronegativity" as the electrostatic force (F) excreted by the effective nuclear charge ($Z_{eff}$) on the valence electrons.

$$\text{that is, } F \mu Z_{eff}/r_{cov} \tag{55}$$

Now the Slater's definition of *IP* is [93]:

$$IP = RZ^2_{eff}/n^{*2}, \tag{56}$$

Substituting $Z_{eff}$ in equation (56) he arrived

$$F \mu n^*(IP/R)^{1/2}/r^2_{cov} \tag{57}$$

Comparing with Pauling's atomic electronegativity value, he proposed the electronegativity ansatz as

$$\chi = 0.241\{n^*(IP/R)^{1/2}/r^2_{cov}\} + 0.775 \tag{58}$$

### *Boyd and Edgecombe's Scale of Electronegativity [94]*

Boyd and Edgecombe [94] proposed an atomic electronegativity scale based on the topological properties of the electron density distributions of molecules, and they extended this method to evaluate group electronegativities.

In this work, Boyd and Edgecombe assumed that there is an electronegativity factor ($F_A$) associated with atom A that is directly proportional to The distance of the bond critical point from the hydrogen atom in AH ($r_H$) and inversely proportional to the electron density at the bond critical point, $p(r_c)$, where $r_c$ denotes the position of the bond critical point. They also observed that this factor fails to allow for the larger size of the heavier atoms.

They define a term "orbital multiplier" $f_{AB}$, as

$$f_{AB} = R_A/(R_A + R_B) \tag{59}$$

where $R_A$ and $R_B$ are the distances from the nuclei to the orbital center.

Boyd and Edgecombe [94] pointed out that the deviation of $f_{AB}$ from 0.5 measures the difference in the electron-attracting power, or electronegativity of the atoms A and B and also discovered that $p(r_c)$ increases monotonically within each period as the atomic number of A, $Z_A$, increases. Thus, $p(r_c)$ increases while $r_H$ decreases. Boyd and Edgecombe [94] assumed that there is an electronegativity factor ($F_A$) associated with atom A that is directly proportional to $r_H$ and inversely proportional to $p(r_c)$. They also assumed that as the electronegativity increases from left to right within each period while $r_H$ decreases, the electronegativity factor varies inversely with the number of valence electrons of the atom A, $N_A$. Boyd and Edgecombe also stated that the electronegativity factor concept fails to allow for the larger size of the heavier atoms.

The term "electronegativity factor" is defined as

$$F_A = r_H/N_A p(r_c)r_{AH} \tag{60}$$

Boyd and Edgecombe [94] expressed the electronegativity of atom A as a power curve of $F_A$

$$\chi_A = aF^b \tag{61}$$

The two constants or parameters are computed as a = 1.938 and b = –0.2502 to provide the electronegativities of Li and F as 1.00 and 4.00 respectively and then Boyd and Edgecombe [94] evaluated the atomic electronegativity of the 21 elements of the second, third and fourth periods using the computed "a" and "b" values and the taking the electronegativities data of Li and F as references.

## Allen's Scale of Electronegativity [13]

Perhaps the simplest definition of electronegativity is that of Allen [13] who stated that electronegativity is the average energy of the valence electrons in a free atom.

Allen proposed the electronegativity ansatz as:

$$\chi_{Allen} = (n_s\varepsilon_s + n_p\varepsilon_p)/(n_s + n_p) \tag{62}$$

where $\varepsilon_s$ and $\varepsilon_p$ are the one-electron energies of s- and p-electrons in the free atom and $n_s$ and $n_p$ are the number of s- and p-electrons in the valence shell respectively. It is usual to apply a scaling factor, $1.75 \times 10^{-3}$ for energies expressed in kilojoules per mole or 0.169 for energies measured in electron volts, to give values which are numerically similar to Pauling electronegativities.

Furthermore, Allen [13, 14] considered electronegativity as configuration energy of the atoms of interest and he stated that "when orbital occupancy is taken into account, it immediately follows that configuration energy (CE), the average one-electron valence shell energy of a ground-state free atom, is the missing third dimension."

For s-p block elements, the Allen's Scale of electronegativity is

$$\chi_{s-p} = (CE)_{s-p} = (n_s\varepsilon_s + n_p\varepsilon_p)/(n_s + n_p) \tag{63}$$

and for the atoms with ground-state configurations $s^nd^m$ and $s^{n-1}d^{m+1}$ , the Allen's Scale of electronegativity is

$$\chi_d = (CE)_d = (p\varepsilon_s + q\varepsilon_d)/(p + q) \tag{64}$$

where $\varepsilon_s$ and $\varepsilon_d$ are the multipulate-averaged one-electron energies of s- and d-orbitals of the atom in the lowest energy configuration respectively. In the free atom n and m are the usual integers such that (p + q) is the maximum oxidation state observed for the atom in any compound or complex ion.

The multipulate-averaged one-electron energies can be directly determined from spectroscopic data, and so the electronegativities calculated by this method are originally referred to as spectroscopic electronegativities by Allen. The credit of the scale is that the necessary data to compute the electronegativities of atoms are available for almost all elements, and hence, this method allows us to compute the electronegativities of the elements which cannot be evaluated by other methods. However, for d- and f-block elements, doubt in the electronic configuration may arise for the calculation of the electronegativity by Allen's method.

### Nagle's Scale of Electronegativity [95]

Nagle's [95] Scale of electronegativity is based on atomic polarizability. The static electric dipole polarizability or simply polarizability is an experimentally measurable property of an isolated atom. The valence electron density is a parameter which can define and measure the electronegativity of an atom. Nagle found that the cube root of this ratio of the number of valence electrons divided by the polarizability, $(n/\alpha)^{1/3}$, can be used as a measure of electronegativity for all s- and p-block elements (except the noble gases). The value fits well with the electronegativities in Pauling Scale and the correlation yields:

$$\chi = 1.66 \ (n/\alpha)1/3 + 0.37 \tag{65}$$

Ghosh and Gupta [51] also proposed a simple relation between $\chi$ and $\alpha$ as:

$$\chi = a(1/\alpha)^{1/3} + b \tag{66}$$

where a and b are two constants for a given period of the periodic table.

### Zheng and Li's Scale of Electronegativity (1990)

Based on the average nuclear potential of the valence electrons, Zheng and Li (1990) discovered a new method for the determination of the effective nuclear charge $Z_{eff}$ and, considering the atomic electronegativity scale of Mulliken, they defined electronegativity as the ratio of $Z_{eff}$ and the mean radius, $<r>_{nl}$ of the outermost electron of an atom:

$$\chi_M = (I + A)/2 = Z_{eff}/ <r>_{nl} \tag{67}$$

### Ghosh's Scale of Electronegativity [50]

Considering the periodic behavior of the electronegativity and the atomic radius, Ghosh [50] put forward a simple equation for evaluating atomic electronegativity as:

$$\chi = a \ (1/r_{abs}) + b \tag{68}$$

where $\chi$ is electronegativity and $r_{abs}$ is the absolute radii of the atoms, a and b are two constants.

### Li and Xue's Scale of Ionic Electronegativity [96]

Li and Xue [96] proposed an electronegativity scale for the elements in different valence states and with the most common coordination number in terms of effective ionic potential. They defined the effective ionic potential as

$$\varphi = n*(I/R)^{1/2}/r \tag{69}$$

where $I_m = R(Z_{eff}/n*)$ is the ultimate ionization energy, n* is the effective principal quantum number and R (in eV) is the Rydberg constant and $r_i$ is the ionic radius.

They proposed that electronegativity of an ion is proportional to the effective ionic potential and proposed a linear relationship between the ionic electronegativity and the effective ionic potential as:

$$\chi_{ion} = a \ \varphi + b \tag{70}$$

where "a" and "b" are the constants.

They evaluated the values for "a" and "b" as a = 0.105 and b = 0.863 through a linear regression of the effective ionic potential with the Pauling electronegativity data.

They calculated the electronegativities of 82 elements in different valence states and with the most common coordination numbers using the above ansatz and found that for a given cation, the electronegativity increases with increasing oxidation state and decreases with increasing coordination number.

It is important here to mention that the Avogadro's Oxygenicity Scale was a crude electronegativity scale, however, a theoretical justification of Avogadro's attempt can be made using the electrification approaches [44, 96] to define electronegativity.

Some important chemical phenomena, such as the ligand field stabilization, the first filling of p orbitals, the transition-metal contraction, and especially the lanthanide contraction, are well-reflected in the relative values of the proposed scale of electronegativity by Li and Xue. The scale can also be used to estimate the Lewis acid strength quantitatively for the main group elements in their highest oxidation state.

### Noorizadeh and Shakerzadeh's Scale of Electronegativity [97]

Parr et al. defined the electrophilicity index, $\omega$ of atoms, ions, and molecules as

$$\omega = \mu^2/2\eta = \chi^2/2\eta \qquad (71)$$

As the electrophilicity [98] of a system is related to both the resistance and the tendency of the system to exchange electron with the environment, Noorizadeh and Shakerzadeh [97] pointed out that the electrophilicity index can be used to measure the electronegativity of the system.

In reference to nucleophilic-electrophilic, acid-base, or donor–acceptor reaction, the electrophilicity index [98] of atoms and molecules seems to be an absolute and fundamental property of such chemical species because it signifies the energy lowering process on soaking electrons from the donors. This tendency of charge soaking and energy lowering must develop from the attraction between the soaked electron density and screened nuclear charge of the atoms and molecules. It, therefore, transpires that the conjoint action of the shell structure and the physical process of screening of nuclear charge of the atoms and molecules lead to the development of the new electrostatic property—the electrophilicity, electronegativity, hardness of atoms and molecules [21, 22, 40, 41].

### Ghosh and Islam's Scale of Electronegativity (2009)

Ghosh and Islam [22] recently pointed out the conceptual commonality between the two fundamental theoretical descriptors, electronegativity and hardness. They concluded that the hardness and the electronegativity originate from the same source, the electron attracting power of the screened nucleus upon valence electrons and discovered the surprising result that if one measures hardness, the electronegativity is simultaneously measured and vice-versa.

They proposed the electronegativity as:

$$\chi = \eta \tag{72}$$

To justify their hypothesis that "if one measures hardness, the electronegativity is simultaneously measured and vice-versa" to compute some descriptors of the real word like the dipole moment, bond distance, reaction surface, and so forth. [17–22].

## COMMON PROPOSITION REGARDING ELECTRONEGATIVITY

A search of literature [25] reveals that a good number of workers converge to number of common proposition regarding electronegativity:

1.  It is a periodic property.
2.  It is an intrinsic atomic property which is associated with shell structure of atoms and arises from the screened nuclear charge.
3.  It is a global property of atoms, molecules, and ions.
4.  It is a property which has to be measure in energy units.
5.  It is an empirical quantity.
6.  Electronegativity is a conceptual entity, not an experimentally observable property.
7.  As there is no quantum mechanical operator for electronegativity, the quantum mechanical evaluation of electronegativity is ruled out [17–20, 22, 40, 41).

There are certain rules for a reasonable scale of electronegativity [14, 48, 99], namely,

1.  Electronegativity scales must have free atom definition.
2.  Electronegativity should be expressed in energy unit.
3.  Contraction of the main transition group elements must be transparent.
4.  Electronegativity values of noble gas elements must be highest in each period.
5.  Electronegativity scales must satisfy the Silicon rule—all metals must have electronegativity values that are less than or equal to that of Si.
6.  Electronegativity scales must satisfy the Carbon rule—the electronegativity value of C has to be greater than, or at least equal to, that of H.
7.  Existence of the metalloid band in the computed electronegativity data of the elements: The six metalloid elements B, Si, Ge, As, Sb, and Te that separate from the non-metals have electronegativity values, which do not allow overlaps between metals and non-metals.
8.  Electronegativity scales should quantify the Van Arkel-Ketelaar triangle.
9.  A high precision is necessary for each scale.
10. In binary compounds, the electronegativity of the constituent atoms clearly quantifies the nature of bonds.
11. Electronegativity scales must be compatible with the elementary quantum concepts such as shell structure, quantum numbers, and energy levels which describe the electronic structure of atoms.

## UNIT OF ELECTRONEGATIVITY

It is very difficult to understand the meaning of a quantity if one does not know its unit properly. The physical picture corresponds to the term "electronegativity" is till now not clear to us. Each scale has its own identity and usefulness in the field of application. They are not comparable to each other, thus the units of different scales are different. In Table 1.1, the units and dimensions of some electronegativity scales are given.

**Table 1.1.** Some scales and their dimensions.

| Scale | Dimension |
| --- | --- |
| Pauling [10] | $(Energy)^{1/2}$ |
| Mulliken [55], Ghosh [50], Ghosh and Gupta [51], Ghosh and Islam [22] | Energy |
| Allred and Rochow [12] | Force |
| Gordy [65], Parr et al. [58] | Energy/electron |
| Sanderson [71] | Dimensionless |
| Allen (13, 14, 2000 ) | Average one-electron energy |
| Walsh [70] | Force/distance |

## INTER-RELATIONSHIP BETWEEN THE ELECTRONEGATIVITY AND OTHER PERIODIC PARAMETERS

It is now well established fact that, like ionization potential, atomic radius, and others electronegativity is also a periodic parameter [22, 25, 50, 51, 100, 101].

When we look in the total periodic table we are convinced that in a period electronegativity would increase monotonically to be maximum at the noble gas elements and in the pattern is repeated next period and for each row, electronegativity would decrease monotonically to be minimum at the last element of each row. The periodic law is a very fundamental law of nature manifested in many physico-chemical properties of atoms and molecules. Although the periodic table does not follow the quantum mechanics, it has chemical organizing power relating many seemingly different properties, which are individually periodic. So, one periodic parameter can be converted to another. We already discussed the various attempts which have been made to evaluate electronegativity of atoms using other periodic parameters, like the atomic radius [12, 50, 65], ionization potential [90], polarizability [51, 94], hardness [22], and so forth. Also there are certain views which suggest a relationship between electronegativity and other periodic parameters. Pearson [102] suggested that for donor atoms, the electronegativity can be taken as a measure of the hardness of the base. After rigorous research on systematic formulation of electronegativity and hardness, Putz [49] opined that the hardness and electronegativity are proportional to each other:

$$\chi \propto \eta \qquad (73)$$

Ayers [103] on the basis of the energy expression of March and White [104] proposed expressions for the electronegativity and the hardness of neutral atoms and pointed out that the two fundamental atomic parameters; hardness and electronegativity are proportional to each other.

Ghosh and Islam [17–20, 22, 40, 41] opined that electronegativity is not an observable property and hence, no quantum mechanical operator can be assembled for its quantum mechanical evaluation. It is an empirical quantity and remains empirical. So, there is a plenty scope of research on this domain. Allen [13, 14] suggested that the concept and scale of electronegativity have a "broken symmetry" symmetry relationship with Periodic Tables categorization, which completes the Periodic Table. Following Pauling, some scientist believe that electronegativity is an in situ property developed on molecule formation rather it is an intrinsic ground state property of atom and it is carried in to molecules but a majority of scientists [13, 14, 22, 25, 48–53], have established that electronegativity is a free atom property. Allen et al. [105, 106] opined that the *in situ* assumption is self defeating and so the electronegativity is very difficult to define.

## CONCLUSION

From the above discussions, it is self evident that no rigorous definition of electronegativity has been suggested so far and the final scale of electronegativity is yet to develop. The problem of unit of electronegativity is probably solved in favor of energy unit. It is also argued that electronegativity is not an *in situ* but an intrinsic free-atom, ground-state property.

The concept of electronegativity and electron attracting power of an atom bonded to divergent atoms are now accepted as true "in each other's pocket." This electron attracting power originates from the effective nuclear charge. It, therefore, transpires that electronegativity is a fundamental property of atomic shell structure and obviously periodic in nature.

Finally, we may conclude that the electronegativity is a fundamental descriptor of atoms molecules and ions which can be used in correlating a vast field of chemical knowledge and experience. During the chemical event of molecule formation, there is a physical process of electronegativity equalization through the rearrangement of charge. The attempts to refine the concept and scale of electronegativity theory are not yet sufficiently complete to enable a judgment to be reached on their effectiveness. We quote original from Pritchard and Skinner,

Meanwhile, it seems safe to say that the chemist will continue to make use of the crude electronegativity theory for some time yet-a practice for which he can hardly be blamed in the absence of an alternative theory of equal generality.

The applications of the concept of electronegativity are an animated field of current research. In Part 2 of this work some of the major applications of electronegativity in the real world of molecular chemistry and molecular biology are reviewed.

## ACKNOWLEDGMENT

We wish to express our sincere thanks to Professor D.C. Ghosh, PRS, PhD, University of Kalyani for his invaluable teaching, discussions and comments on this topic.

**KEYWORDS**

- **Electronegativity**
- **Heteronuclear**
- **Nucleophilic–electrophilic**
- **Orbital electronegativity**
- **Oxygenicity scale**
- **Polarizability**
- **Quantum thermodynamic**

# Chapter 2

## The Time Evolution of the Electronegativity Part-2: Applications

Nazmul Islam and Chandra Chur Ghosh

---

### INTRODUCTION

The improvement of chemistry is often based on some empirical relations between some properties, which can be measured or can be found in literature and those which one wants to investigate. This is thoroughly useful because both quantities result from the same electronic structure of the molecule under consideration. The ultimate aim of chemists, till now, is to solve the fundamental and naturally occurring phenomena. For this purpose, chemists relying upon the electronic structure of the fundamental structural construct of the universe—the atom and based upon some natural occurring phenomenon and experimental observations introduced some very important structural principles. Unfortunately, many of such things are not the things of the real world. The concepts which do not originate from a "strong" theory but clearly and vividly describe a series of relations among chemical data are called the empirical entities. They are very important for the rationalization and prediction of various physico-chemical phenomena. Because of the simplicity the empirical entities accompany our structural thinking in terms of the scientific languages. Sometimes the empirical entities can be improved on the bases of some theoretical methods. As a result, time to time new concepts have been introduced in chemistry for the explanation and prediction of several observable facts, the properties of atoms and molecules and also the reactions. The theory approaches a chemical experiment via some selective approximations and simplifications which then serve as bridge between the rigorous theory and chemical reality. In 1939, in the first edition of *The Nature of the Chemical Bond*, Pauling [1, 2] introduced an important fundamental descriptor of science—the electronegativity. Pauling defined electronegativity as "the power of an atom in a molecule to attract electrons to it." Till then, attempts are made to define and quantify the hypothetical entity—the electronegativity, but the basic tenet of Pauling—"the power of atoms to attract or retain or not to release electrons" is not yet changed. There is a maxim that when there are many treatments for a disease, none of them is completely adequate. The same idea could be applied to electronegativity in view of the many attempts to define and quantify it. Allen [1989, 3] was the first who recognize that the electronegativity concept and its scale can indeed furnish the appropriate function to reproduce the observed periodicities of experimental quantities.

In the Part-1 of this review work, we [4] have analyzed the fact that exact definition and the best method for the evaluation of electronegativity remains to be discovered. There is plenty of room for controversy, and no two workers will agree completely.

Electronegativity is not an observable property and hence, no quantum mechanical operator can be constructed for its quantum mechanical evaluation. The appearance and significance of heuristically developed concepts of electronegativity resemble the unicorns of mythical saga (Frenking and Krapp, 2007). They exist but never seen. Without the concept and operational significance of electronegativity, chemistry, and many aspects of condensed matter physics becomes chaotic and the long established unique order in chemico-physical world will be disturbed.

However, there is no experimental benchmark for the electronegativity, the concept of electronegativity is one of the most useful theoretical descriptor which is widely used by chemists, physicists, engineers, biologists, and geologists. Some very fundamental conceptual descriptors such as bond energies, bond polarities, and the dipole moments, force constants, and the inductive effects can only conceived in terms of the concept of electronegativity [5–7].

The significance of the electronegativity concept in chemistry become transparent from the remarks of Coulson [8] *"the astonishing success which the theory has had in correlating a vast field of chemical knowledge and experience."* It is also opined (Jensen, 1996) that no concept more thoroughly encompasses the fabric of modern chemistry that that of electronegativity.

Various scientists reviewed the concepts and the applications of electronegativity [7, 9, 10]. In this work, we will pinpoint some of the *"astonishing successes"* of the electronegativity concept in the real world and its' relationship with other periodic parameters in the following sections.

## THE ELECTRONEGATIVITY EQUALIZATION PRINCIPLE

Understanding chemistry requires a basic understanding of the structure of atoms, ions, and molecules and their interactions. After the announcement of the electronegativity equalization principle [11], it becomes one of the most useful applications of the electronegativity. Focus on the chemical bond that holds together the atoms forming the molecule, an answer of the fundamental question—"why do atoms interact to form molecule?" was given by the electronegativity equalization principle. Sanderson [11] stated that when two atoms having different electronegativities come together to form a molecule, the electronegativities of the constituent atoms become equal, yielding the molecular equalized electronegativity. Thus, for the first time, the concept of electronegativity had been taken as a dynamic property rather than a static one. Electrons in a stable homonuclear covalent bond are equally attracted to both nuclei. But this is not true for the heteronuclear system, where two atoms (or more) having different electronegativity values are joined through covalent bond. The more electronegative atom having more electron attracting power attracts the bonding electron pair more towards itself. Thus some amount of charge transferred from the lower electronegative atom to the higher one. This can also be viewed as charge is transferred from the atom having higher chemical potential value to the atom having lower chemical potential value until both the chemical potential and electronegativity of the constituents becomes equal.

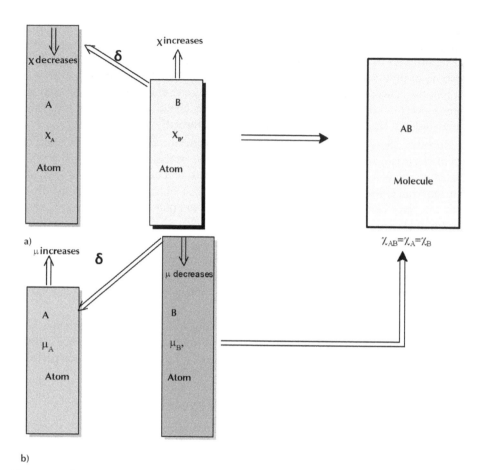

**Figure 2.1.** Electronegativity (chemical potential) equalization scheme.
where $\chi_A > \chi_B$
(a) Electronegativity equalization
(b) Chemical potential equalization

Two different electronegative atoms have atomic orbitals of different energies. The process of bond formation must provide a pathway by which the energies of the bonding orbitals become equalized. If in the bond formation process the electronegativity of the higher electronegative atom decrease as that atom acquires electronic charge ($\delta$) and that of lower electronegative atom increase as it loses the electronic charge ($\delta$). Sanderson [11] postulated a geometric mean principle for the electronegativity equalization. He pointed out that the final electronegativity of a molecule is the geometric mean of the original atomic electronegativities. The electronegativity equalization principle is now linked to the fundamental quantum mechanical variation principle. Parr, Donnelly, Levy, and Palke [12] identified electronegativity as the amount of energy required to remove a small amount of electron density from the molecule at the point r, that is,

$$\chi(r) = \delta Ev(\rho)/\delta\rho(r) \qquad (1)$$

Parr et al. [12] noted that the energy is minimized only if the electronegativity is equalized, because if there are two place in the molecule with different electronegativity, then transferring a small amount of electron density q, from the atom having lower electronegativity to the atom with greater electronegativity will lower the energy. Parr and Bartolotti [13] gave a proof of the electronegativity equalization principle from a sound density functional theoretical [14, 15] background. The term "chemical potential" as it occurs in thermodynamics [16] has long been accepted as a perspicuous description of the escaping tendency of a component from a phase. Parr et al. [12] identified electronegativity as the negative of the chemical potential of the system. They also pointed out that both parameters can be adopted at the molecular level because they have the very same properties in the charge equalization procedure. Parr et al. [12] correlated charge transfer, the electronegativity difference, and the energetic effect of the charge transfer with the geometric mean principle of electronegativity equalization [11].

Let us consider the formation of a molecule AB, in its ground state, from the constituent ground state gaseous atoms A and B have the chemical potentials $\mu_{AB}$, $\mu^0_A$, and $\mu^0_B$, the electron densities $\rho_{AB}$, $\rho^0_A$, and $\rho^0_B$, the numbers of electrons N, $N^0_A$, and $N^0_B$, and the nuclear potentials $\upsilon_{AB}$, $\upsilon^0_A$, and $\upsilon^0_B$ respectively. Of course, the chemical potentials, $\upsilon_{AB}$ of the product molecule is $\upsilon_{AB} = \upsilon^0_A + \upsilon^0_B$, and the number of electrons $N = N^0_A + N^0_B$.

The number of electrons which flow from B to A during the formation of AB molecule is given [13] as:

$$\Delta N = (\tfrac{1}{2}\,\gamma)\ln(\mu^0_B/\mu^0_A) \qquad (2)$$

The $\gamma$ is not always constant rather it changes in a fairly narrow range of 2.15 ± 0.59 [13].

The energy difference $\Delta E$, is correlated with the standard chemical potential difference of the atoms A and B ($\mu^0_A - \mu^0_B$) and the number of electron transferred $\Delta N$ as follows:

$$\Delta E = (\mu^0_A - \mu^0_B)\,\Delta N \qquad (3)$$

Ray, Samuels, and Parr [17] first derived the necessary algorithms for the equalized molecular electronegativity and other descriptors such as bond distance, force constants, and so forth using the Simple Bond Charge (SBC) model [18–21]. For a diatomic molecule AB with the equilibrium bond length $R_{AB}$ if we consider $Z_A$ and $Z_B$ as the charge on atom A and B in the diatomic molecule AA having the bond length $2r_A$ and the charge on B in BB having the bond length $2r_B$ respectively and δ is the amount of charge transferred during the process of the molecule formation, then ($Z_A + δ$) and ($Z_B - δ$) will be the charges on nuclei A and B in the molecule AB.

Pasternak [21] defined the electronegativity of a bonded atom A in a molecule as:

$$\chi_A = C(Z_A/r_A) \qquad (4)$$

where C is the constant which depends on the nature of the bonding between A and B.

Thus, the electronegativity of atoms A and B in the molecule has the form:

$$\chi_A(\text{in AB}) = C(Z_A + \delta)/r_1 \tag{5}$$

and,

$$\chi_B(\text{in AB}) = C(Z_B - \delta)/r_2 \tag{6}$$

According to Sanderson's electronegativity equalization principle [11], electronegativity of bonded atoms in a molecule must be equal to each other. Thus,

$$\chi_{AB} = \chi_A(\text{in AB}) = \chi_B(\text{in AB}) = C(Z_A + \delta)/r_1 = C(Z_B - \delta)/r_2 = C(Z_A + Z_B)/R_{AB} \tag{7}$$

$$\text{or, } \chi_{AB} = (\chi_A r_A + \chi_B r_B)/(r_1 + r_2) = (\chi_A R_{AA} + \chi_B R_{BB})/2R_{AB} \tag{8}$$

Huheey [22] pointed out that the assumption of electronegativity equalization ignores energies arising from electrostatic (ion-ion) interaction [23] and changes in overlap [24]. These are true for extremely ionic bonds, but for the bonds which are predominantly covalent in nature the errors incurred are small. The errors resulting from neglect of changes in electrostatic and overlap terms have opposing effects and tend to cancel each other; both approach zero as δ approaches zero.

## JUSTIFICATION OF THE REACTION SURFACE IN TERMS OF ELECTRONEGATIVITY

Sekhon [25] demonstrated the use of the equalized electronegativity as a guide to the outcome of metathesis reactions in inorganic and organic chemistry. He had examined some examples of the exchange reactions of the type AB + CD = AD + BC and concluded that an exchange reaction proceeds from the left to right if the total sum of the equalized electronegativity of the products is greater than that of the reactants, that is,

$$-\Delta E_{Reac} \approx \tfrac{1}{2}\left\{\sum_{i=1}^{n} (\chi_{Product})_i - \sum_{j=1}^{m} (\chi_{reactant})_j\right\} \tag{9}$$

Or simply,

$$-\Delta E_{Reac} = \Delta\chi \tag{10}$$

where n and m are the number of products and reactants respectively, and

$$\Delta\chi = \tfrac{1}{2}\left\{\sum_{i=1}^{n} (\chi_{Product})_i - \sum_{j=1}^{m} (\chi_{reactant})_j\right\} \tag{11}$$

## ELECTRONEGATIVITY AND MOLECULAR ORBITAL THEORY

Coulson [26] introduced the Linear Combination of Atomic Orbital (LCAO) approximation and proposed the LCAO-MO theory which introduced two important quantities:

1.  the Coulomb integral, $\qquad \alpha_r = \int \Phi_r \hat{H} \, \Phi_r \, d\tau,$ $\qquad\qquad$ (12)

The Coulomb integral $\alpha_r$, measures the energy of an electron when confined to the atom r within the molecule

and,

2. the resonance integral,    $\beta_{rs} = \int \Phi_r \hat{H} \, \Phi_s \, d\tau.$    (13)

Coulson and Longuet-Higgnhs [27] point out that the Coulomb term can be related to the ionization potential (I) of atom r. But after a detail examination on this topic, Mulliken [28, 1949] concluded that in a bond XY the difference, $a_X - a_Y$, should be proportional to the difference in electronegativity of the atoms X and Y in their proper valence states.

## THE DIPOLE CHARGE AND DIPOLE MOMENT IN TERMS OF ELECTRONEGATIVITY

The most obvious application of electronegativities is the prediction of the polarity of a chemical bond, for which the concept was originally introduced by Pauling. In general, the greater is the difference in electronegativity between two atoms, the more polar is the bond that will be formed between them, with the atom having the higher electronegativity being at the negative end of the dipole. Dipole is an index or descriptor of the asymmetry of charge distribution in molecules. The charge distribution pattern in the hetero nuclear diatomic molecules always generates dipole. The bond character is a common topic in chemistry for the determination of the physical and chemical behavior of compounds. The simplest way to determine bond character is to use the electronegativity difference between the bonded atoms. Again the dipole moment $\mu$ of a diatomic molecule AB has been related to the difference of the two atomic electronegativities $(\chi_B - \chi_A)$.

According to Mulliken [29] and Dailey and Townes [30] the dipole moment in hetero nuclear diatomic molecules has several components that is,

$$\mu = \mu_{atomic} + \mu_{overlap} + \mu_{hybridization} + \mu_{polarization} \qquad (14)$$

However, the principal components are two. The first one is bond moment which is developed due to unequal sharing of charge between nuclei goaded by their difference in electronegativities. The second component originates from the asymmetry of charge distribution in lone pair electrons in hybrid orbitals.

$$\mu = \mu_{atomic} + \mu_{hybridization} \qquad (15)$$

Mulliken [29] pointed out that, in diatomic molecules, the bond pair moment is the main contributor to the electronic dipole moment of the molecules. However, the two contributing components of dipole are deeply interlinked with the pattern of electron distribution in molecules. Coulson (1961) and Dewar (1969) suggested the fact that the lone pair of electrons accommodated in a hybrid orbital acquires asymmetry of charge density distribution and generates atomic dipole. Ghosh and Bhattacharyya [31] have derived quantum mechanical algorithm of dipole moment of molecules.

Between these two components of the dipole moment, we can calculate the bond moment part semi-empirically but the lone pair contribution to the dipole moment of molecules can only be calculated quantum mechanically and the possibility of empirical evaluation of lone pair component of dipole is ruled out.

Dipole is a descriptor of the asymmetry of charge distribution. Let us reproduce the quantum mechanical algorithm developed by Ghosh and Bhattacharya [31].

The permanent quantum mechanical electric dipole moment, $\mu$ of the molecule whose electric state is given by $\Psi_{el}$ is

$$\mu = \int \Psi^*_{el} \, d_{op} \, \Psi_{el} \, d\tau \tag{16}$$

where $d_{op}$ is the quantum mechanical operator of dipole moment. The electric dipole moment operator, $d_{op}$, for a molecule includes summation over both the electronic and nuclear charges.

$$d_{op} = \sum_i (-er_i) + \sum_\alpha Z_\alpha er_\alpha \tag{17}$$

where $r_\alpha$ is the vector from the origin to the nucleus of atomic number $Z_\alpha$ and $r_i$ is the vector to the electron i.

Since the second term in equation (17) is independent of the electronic coordinates, we have

$$\mu = \int \psi_{el} \{\sum_i (-er_i)\} \psi_{el} d\tau + \sum_\alpha Z_\alpha er_\alpha \int \psi^*_{el} \psi_{el} d\tau \tag{18}$$

or,

$$\mu = -e \int \psi^2_{el} \sum_\alpha Z_\alpha r_\alpha \tag{19}$$

We can also rewrite this expression as

$$\mu = -eN \int \psi^2_{el} \sum_i r_i d\tau + e \sum_\alpha Z_\alpha r_\alpha \tag{20}$$

where N is the number of electrons in the molecule and $r_i$ is the position vector of electron i. or,

$$\mu = -e \int\int\int \rho(x, y, z) \, r \, dxdydz + e \sum_\alpha Z_\alpha r_\alpha e \tag{21}$$

where $\rho(x, y, z)$ is the electronic probability density.

Now expanding $\rho$ in terms of the molecular orbitals and then expanding the molecular orbitals, in turn, in terms of the atomic orbitals according to the LCAO-MO SCF scheme and invoking the necessary approximations of Pople's method [32, 33], the molecular dipole moments are obtained as a sum of two components.

$$\mu_{total} = \mu_{at} + \mu_{sp} \tag{22}$$

The first term arises from the contribution of net atomic charge densities and second one is the contribution of atomic dipoles resulting from the mixing (hybridization) of s and p orbitals on the same atom center. This method has been applied to solve many puzzles of chemistry. We mention one important application of the method of dipole correlation of electronic structure is in resolving the mystery—whether the two lone pairs of electrons of O of water molecule is the same (squirrel ear) or not? [34].

Pauling [1, 2] first assumed that the ionic character may be obtained from the dipole moment of a compound and he compared the dipole moment with the moment produced by the two point charge at a distance apart equal to the internuclear separation of the molecule.

that is,                                                                         q $\mu\mu$                                            (23)

The working formulae to evaluate the dipole charge are reviewed below:

## Pauling's Formula

Pauling, in an attempt to evaluate the dipole charge, plotted the percentage of ionic charges against their electronegativity differences. Pauling [2] proposed an ansatz to calculate the ionic character of the bond (i.e., static charge) was as

$$q = 1 - \exp(-(\chi_B - \chi_A)^2/4)$$                                    (24)

where $\chi_B$ and $\chi_A$ are the atomic electronegativities of atoms B and A respectively.

## Nethercot's Formulae

Nethercot [35, 36] concluded that q could not be a simple function of electronegativity difference of two atoms and he proposed two formulae to calculate the dipole moment charges. His proposed ansatz(s) were

$$q = 1 - \exp(-3(\chi_B - \chi_A)^2/2\,\chi_{AM}^2)$$                      (25)
$$q = 1 - \exp(-(\chi_B - \chi_A)^{3/2}/\chi_{GM}^{3/2})$$                 (26)

where $\chi_{AM}$ and $\chi_{GM}$ are the arithmetic mean (AM) and the geometric mean (GM) of the two atomic electronegativities.

## Barbe's Formula

Barbe [37] proposed a simple equation to calculate the dipole moment charges of hetero nuclear diatomic molecules as:

$$q = (\chi_B - \chi_A)/\chi_B$$                                          (27)
$$\text{or, } q = \Delta\chi/\chi_B$$                                    (28)

where $\chi_B > \chi_A$.

## COMPUTATION OF BOND MOMENT

The electric dipole moment is a measure of the separation of positive and negative electrical charges in a system of charges that is a measure of the charge system's overall polarity. In the simple case of two point charges, one with charge +q and one with charge –q, the electric dipole moment is

$$\mu = q \times d$$                                                     (29)

where d is the displacement vector and $\mu$ is the electric dipole moment vector generated by bond charge.

If we take a series of di-atomics whose bond distances, d can be known to a satisfactory accuracy. So, the dipoles could be calculated if the q's are known. Let us recast the equation (29) to calculate the dipole in Debye unit.

$$\mu = q\, e\, R_e$$                                                    (30)

where $\mu$ is the dipole moment of molecules in debye unit, q is the dipole moment charge on atomic site, e represents the electronic charge in esu unit and $R_e$ is the internuclear distance of diatomic molecules in centimeter unit.

As electronegativity is an abstractly defined property and also it is not an observable, hence it cannot be directly measured. However, relative electronegativities can be observed indirectly by measuring molecular dipole moments: in general, the greater the dipole moment, the greater the separation of charges must be, and therefore, the less equal the sharing of the bonding electrons must be.

It has long been recognized that the dipole moment, $\mu_d$, of the molecule AB can be related to the difference of the two atomic electronegativities $(\chi_B \sim \chi_A)$. Indeed in 1932, Pauling [1, 2] proposed the empirical relationship $q = 1 - \exp[-(\chi_B - \chi_A)^2]$ for estimating the ionic character in the molecule. This form was chosen to agree with the then available experimental values of the "dipole moment" charge. In the next year, Malone [38] discovered that dipole moment in Debye $(\mu_d)$ of a hetero nuclear bond A-B and the electronegativity difference, $\chi_A - \chi_B$, are proportional, that is,

$$\chi_A \sim \chi_B \mu \mu_d \tag{31}$$

In 1935, Mulliken [29] in an attempt to correlate the relative electronegativities, effective charges on atoms in partially polar molecule, dipole moments in terms of LCAO-MO coefficients, proposed that the coefficients are affected by the polarities of the bond and hence are important for the prediction of the electronegativity.

The bonding molecular orbital $\Phi_{AB}$, can be written as:

$$\Phi_{AB} = a\Phi_A + b\Phi_B \tag{32}$$

If a = b, then the two atoms are same electronegativity, but if a > b then we can say that electronegativity of A > B. Thus, Mulliken [29] pointed out that the difference of electronegativity can be correlated with the difference of the LCAO coefficients, that is, a – b or more precisely $a^2 - b^2$.

For a diatomic molecule where the bonding electron pair only contributes to the electrical moment the electronic distribution occurs as:

- $2ea^2$ is the charge centered on the atom A,
- $2eb^2$ is the charge centered on atom B.
- 4eabS is the charge centered the centers A and B and the distance of the electric center from the midpoint of the bond A-B is considered as z. Then the electrical net moment can be obtained by the contribution of the charges centered in centers A, B, in between A-B and a charge + e on each atom.

$$\mu_{Net} = er(a^2 - b^2) - 4ezabS \tag{33}$$

where r is the distance between the atom A and B.

Mulliken [29] defined the term z as the measure of inequality of the polarity of the two atoms A and B. If the molecule AB is homopolar then z = 0 its magnitude increases with the inequality of the size of A and B. its sign is such that its positive pole is directed towards the larger atom. Mulliken empirically correlated the charge distribution of the molecule in terms of the electronegativity and proceeds to evaluated the primary dipole moment of the molecule in terms of the LCAO coefficient as:

$$(\mu_d)_{AB} = Q_B \, re - (4ezabS - \mu_s) \tag{34}$$

where Q is the net charge and $Q_A = -Q_B = -e(a^2 - b^2)$. The term "4ezabS" is the homopolar dipole and "$Q_B \, re$" is the main dipole term.

In case of heteropolar system, this term is cancelled by $\mu_s$.

Thus for the heteropolar system, we can write

$$(\mu_d)_{AB} = Q_B \, r_e \tag{36}$$

Thereafter, in 1954, Dailey and Townes (1954) in terms of LCAO coefficients (a and b) proposed that the dipole moment for a heteropolar bond can be expressed as sum of the primary moment ($\mu_p$), overlap moment ($\mu_o$) of the orbitals of atoms, hybridization moment ($\mu_{hy}$) of the valence shell of the atoms and polarization moment ($\mu_p$) of the non bonding electron.

$$\mu_d = \mu_p + \mu_o + \mu_{hy} + \mu_p \tag{37}$$

They defined the ionic character of a heteropolar bond as the difference between polarizabilities for the electron to be found on atom A or B. They detonated it in terms of the LCAO coefficient as $(b^2 \sim a^2)$

$$\mu_d = eR(b^2 \sim a^2) + \mu_o + \mu_{hy} + \mu_p \tag{38}$$

Dailey and Townes (1954) also pointed out that it is not possible to calculate the contribution of the polarization of the non bonding electron to the total dipole moment and considered the earlier assumption (Pauling) that the first term of the equation is of major importance.

that is,           $$\mu_d = eR(b^2 \sim a^2) \tag{39}$$

## COMPUTATION OF HETERO POLAR BOND LENGTH IN TERMS OF ELECTRONEGATIVITY

Pauling [1, 2] first evaluated the bond length from the electronegativity. The Pauling electronegativity was derived from heats of formation or essentially, bond energies, the electronegativity difference between two atoms reflects the strength of two bonds and moreover there exist a quantitative correlation between electronegativity and bond polarity. Now, let us consider the formation of a diatomic molecule AB from its constituent atoms A and B as follows,

$$A + B \rightarrow AB \tag{40}$$

Let the equilibrium bond length, the electronegativity of the molecule AB are $r_{AB}$, $\chi_{AB}$, and the electronegativities of atoms A and B are $\chi_A$ and $\chi_B$ respectively. Now let us imagine that, after the formation of the molecule, a point charge is located at a distance $r_1$ from A and $r_2$ from B with the condition $r_1 + r_2 = r_{AB}$. When atoms approach to form the molecule, the electron density function over the whole space undergoes rearrangement. Thus, there is a physical process of inter atomic charge transfer and rearrangement during the chemical event of the formation of hetero nuclear molecules goaded by the physical process of electronegativity equalization. Let the electronegativities of the atom A and B in the molecule AB are $\chi/_A$ and $\chi/_B$ respectively. The principle of electronegativity equalization provides,

$$\chi_{AB} = \chi/_A = \chi/_B \tag{41}$$

Now, on the basis of SBC model, Ray et al. [17] derived the internuclear bond distances of hetero nuclear diatomics using the concept of electronegativity equalization and the zero order approximation of Pasternak [21], that is, $r_A = r_1$ and $r_B = r_2$. We reproduce below the process of evaluation of inter nuclear equilibrium bond distance of Ray et al. [17].

$$R_{AB} = (r_A + r_B) - \{(r_A r_B (\chi^{1/2}_A - \chi^{1/2}_B)^2\}/(\chi_A r_A + \chi_B r_B) \tag{42}$$

where $r_A$ and $r_B$ are the covalent radii of the atom A and B respectively.

In a recent work, Ghosh and Islam [39–43] have modified the expression for computing the inter nuclear bond distances in terms of electronegativity and size data of atoms by substituting the covalent radii by most probable radii or absolute radii of atoms in the above equation The modified expression of $R_{AB}$ is:

$$R_{AB} = (r_A' + r_B') - \{(r_A' r_B' (\chi^{1/2}_A - \chi^{1/2}_B)^2\}/(\chi_A r_A' + \chi_B r_B') \tag{43}$$

where $r_A'$ and $r_B'$ are the most probable radii or absolute radii of the atoms A and B respectively.

## ATOMIC POLAR TENSOR

Kim [44] extended the SBC model to evaluate the Atomic polar tensor. He showed that electronegativity and the electronegativity equalization can be used as an important tool for the determination of Atomic Polar Tensor in case of diatomic molecule.

Kim [44] evaluated the dipole charge as:

$$q = \{ r_1 r_2 / CR_{AB} \}(\chi_B - \chi_A) \tag{44}$$

The centroid of positive charge r, relative to the point defining the centroid of negative charge is given as:

$$r = \{r_2 Z_B - r_1 Z_A - (r_1 + r_2) q\}/(Z_A + Z_B) \tag{45}$$

Kim [44] defined the dipole moment, $\mu$, as:

$$\mu = (Z_A + Z_B)r = -(r_1 + r_2) q + (r_1 Z_B - r_2 Z_A)$$
$$= - R_{AB} q + (r_2 Z_B - r_1 Z_A)$$
$$= - 1/C [(r_A r_B \chi_A \chi_B)/(r_A \chi_A + r_B \chi_B)^2][R^2_{AB} (\chi_B - \chi_A)] \tag{46}$$

For AB type diatomic molecule, where A atom is located at the origin and B atom in positive Cartesian direction and $\chi_B < \chi_A$ the atomic polar tensors ($P_x$'s) for atoms. A and B was given as:

$$P_x^B = - P_x^A = (\partial\mu/\partial R)_e \tag{47}$$

where $(\partial\mu/\partial R)_e$ is the dipole moment derivative at geometric equilibrium.

Differentiation of the equation (47) with respect to R gives the atomic polar tensor of B atom:

$$P_x^B = (\partial\mu/\partial R)_e = -(\chi_B - \chi_A) . 2R_{AB}r_A r_B \chi_A \chi_B/C(r_A\chi_A + r_B \chi_A)^2 \qquad (48)$$

## BOND STRETCHING FREQUENCY AND FORCE CONSTANT

Several correlations have been shown between infrared stretching frequencies of certain bonds and the electronegativities of the atoms involved however, this is not surprising as such stretching frequencies depend in part on bond strength, which enters into the calculation of Pauling (1987) electronegativities. The most commonly encountered form of Hooke's law is probably the spring equation, which relates the force exerted by a spring to the distance it is stretched by a force constant, k, measured in force per length.

$$F = -kx \qquad (49)$$

with,

$$k = -W_1/R^3 \qquad (50)$$

Now, let us discuss the algorithm of Ray et al. [17] for the computation of the force constant on the basis of SBC model. In this model, the vibrational energy function was defined as:

$$W = W_0 + (W_1/R) + (W_2/R^2) \qquad (51)$$

$$W_1/R = -(Z_A + Z_B)[\{(Z_A + \delta)/r_1\} + \{(Z_B - \delta)/r_2\} - \{(Z_A + \delta)(Z_A + \delta)/R(Z_A + Z_B)\}] \qquad (52)$$

and

$$W_2/R^2 = h^2(Z_A + Z_B)/8mR^2\upsilon^2_{AB} \qquad (53)$$

where W is the Born–Oppenheimer potential and R is the internuclear distance between A and B. The term, $W_1/R$ and $W_2/R^2$ are assigned to describe the electrostatic energy of the system and the kinetic energy of the bond charge moving freely in a one-dimension box of length $\upsilon R$ along the bond respectively.

Using the equalized molecular electronegativity expression, $\chi_{AB} = (\chi_A R_{AA} + \chi_B R_{BB})/2R_{AB}$, Ray et al. [17] obtained:

$$W_1/R = -\{(Z_A + Z_B)^2/R\}[2 - \{r_1 r_2/(r_1 + r_2)^2\}] \qquad (54)$$

Ray et al. [17] found that the quantity in bracket is close to 7/4 for most of the reasonable values of $r_1$ and $r_2$, thus they set it equal to 7/4 to obtain a formula having maximum simplicity. Thereafter using the force constant formula, $k = -W_1/R^3$, they defined the force constant as:

$$k = (7/4)\chi_{AB}^2/(R_{AB}C^2) \qquad (55)$$

where C is the constant which depends on the nature of the bond between A and B.

Badger [45] correlated equilibrium bond distance (R) and the bond stretching force constant (k) as:

$$k = (c/R)^{1/2} + d, \tag{56}$$

where c and d are the constants.

Remick [46] pointed out that the bond stretching force constant (k) is a function of the product of the electronegativities of the atoms constituting the molecule. Thus for a diatomic molecule, the bond stretching force constant ($k_{AB}$) was given by:

$$k_{AB} = f(\chi_A \chi_B) \tag{57}$$

The stretching force constant is not only functions of product of the electronegativity of the atoms present on the molecule but it is also dependent on several other factors such as (i) bond order (N), and (ii) bond distance (R).

Gordy [47] considered all the factors and modified the Reimick's equation as:

$$k_{AB} = f(N, \chi_A \chi_B, R) \tag{58}$$

Thus, he proposed the general equation for the bond stretching force constant as follows:

$$k_{AB} = a\, N(\chi_A \chi_B / R^2)^{3/4} + b, \tag{59}$$

where a and b are constants for certain broad classes of molecules. To determine the constants, a and b, Gordy compared calculated $N(\chi_A \chi_B / R^2)^{3/4}$ with the experimental bond stretching values of some molecules and evaluated the constants as a = 1.67 and b = 0.30.

It should be notated here that $k_{AB}$ is measured in dynes × cm $10^{-5}$, R is in Å. The value of the constants "a" and "b" varies from a class of molecule to another class. A detail study was made by Gordy and proposed different constant values for different classes and bond order in his paper [47].

On the basis of Pasternak's [21] electronegativity definition and SBC model [18–21], Ray et al. [17] proposed a direct method for the evaluation of the force constant of A-B bond from the equalized electronegativity as:

$$k_{AB} = (7/4)\, \chi^2_{AB} / R_{AB} C^2 \tag{60}$$

where, C is a constant which depends on bond type.

## STANDARD ENTHALPIES OF FORMATION AND BOND DISSOCIATION ENERGY

Bratsch [48] pointed out that the Pauling Scale of electronegativity can be used to predict standard enthalpies of formation of binary compounds:

$$\Delta H^0_f \text{ (kilo joule/mole)} = -96.5n\, [\chi_A - \chi_B]^2 \tag{61}$$

where n is the number of equivalents in the compound formula, $\chi_A$, $\chi_B$ are the Pauling electronegativity in $(eV)^{1/2}$ unit.

Knowledge of bond dissociation energies of chemical bonds in molecules is essential for understanding chemical processes [49]. At the outset, it should be emphasized that there is a distinction [50] between the bond dissociation energy and the average bond energy. The bond dissociation energy of a bond in a molecule A-B is the energy needed to separate the radicals A and B to infinity, each species being in its ground state. The average bond energy of a bond A-B is defined as $1/n^{th}$ the energy needed to separate each of the atoms in a symmetrical molecule AB, to infinity, all species being in their ground states, that is, $1/n^{th}$ of the heat of atomization of the molecule. In general, the bond dissociation energy plays a more important role in chemistry than the average bond energy. Szwarc et al. [50] have demonstrated the importance of bond dissociation energies in the interpretation of chemical kinetic data. Szwarc et al. [50] have shown how the analysis of complex reactions into their elementary reactions can be aided greatly by the knowledge of the bond dissociation energies. Because thermal, photochemical, radiation, and discharge reactions are usually complex, tables of bond dissociation energies should be very useful in interpreting kinetic data in these fields. Bond dissociation energies also can be employed to calculate thermochemical properties. Heats of formation of radicals can be obtained from the bond dissociation energies, which, in turn, can be employed for calculating heats of reactions involving free radicals. Furthermore, bond dissociation energies can be used to calculate heats of reactions in which free radicals are not involved.

Finally, knowledge of bond dissociation energies is essential in interpreting most results obtained by electron bombardment of molecules. In a mass spectrometer one measures the appearance potential of an ion of known mass. In order to deduce what process occurred in the mass spectrometer, one calculates the appearance potentials of various possible processes and by comparing those to the observed value the most likely process can be chosen. It is opined that the knowledge of the bond dissociation energies is very essential to calculate the appearance potentials of various possible processes (more on the topic please see [51]).

The bond dissociation energies or the bond energy (BE) for a bond A-B can be define as the standard-state enthalpy change for a reaction of the type AB $\rightarrow$ A + B at a specified temperature (T).

$$(BE)_T = \Delta H_f^0(A) + \Delta H_f^0(A) - \Delta H_f^0(AB) \qquad (62)$$

where $\Delta H_f^0$ is the standard state heat of formation.

The earliest method which correlates the BE with electronegativity was the landmark Pauling's equation [1, 2]. The scale correlates the extra ionic resonance energy to the electronegativities of atoms. The energy difference $\Delta$, was defined by:

$$\Delta = D(A - B) - (1/2)[D(A - A) + D(B - B)]. \qquad (63)$$

Pauling was able to assign electronegativity of many elements which roughly satisfy the equation

$$\Delta = (\chi_A - \chi_B)^2 \qquad (64)$$

Pauling took $\Delta = -\Delta H_f^0$ (heat of formation).

Within the framework of SBC model [18–21], Pasternak [21] pointed out that in the limit of infinite internuclear separation, a diatomic molecule becomes two ions at a infinite separation and a bond charge at an infinite distance from the ions and free to extend over an infinite volume. The energy of this system from the equation $W = W_0 + (W_1/R) + (W_2/R^2)$ is $W(\infty) = W(0)$, but the $W(0)$ value is arbitrary. The real end product of the dissociation of a homonuclear diatomic molecule is of course two neutral atoms at infinite separation. Therefore, having the $R = \infty$ configuration of the ions and the bond charge, Pasternak then divided the bond charge and recombined it with the ions. This involves the gain in energy by an amount $W_i$, equal to the twice of the energy required to removal of $Z$ electrons from a single atom.

Thus, the energy of the system of two infinitely separated systems of two infinitely separated neutral atoms is zero.

Thus,
$$W(0) = W_i \tag{65}$$

The dissociation energy D is:

$$D = W(\infty) - W_i - W(R) \tag{66}$$

Or,
$$D = -W_i - (W_1/R) - (W_2/R^2) \tag{67}$$

Pasternak [21] established that the term $W_i$ is a continuous function of $Z$ and can be written as:

$$W_i = 2(aZ + bZ^2) \tag{68}$$

where the coefficients a and b are determined from the experimental ionization energies.

Although many bond dissociation energies of chemical bonds between various kinds of atoms in many molecules have been determined experimentally, there is at present no reliable theoretical or empirical method for estimating bond dissociation energies.

## STABILITY RATIO

The ratio of the average electronic density of an atom to that of a hypothetical, isoelectronic inert atom, is termed (Sanderson, 1952) as the "stability ratio" (SR)

The term SR is defined (Sanderson, 1952) as:

$$SR = D/D_i \tag{69}$$

with
$$D = 3Z/4\Pi r^3 \tag{70}$$

$D_i$ was defined as the electron density of an isoelectric inert atom, determined by interpolation between real values, which was needed to correct the average electron density (D) for variations in $Z$ that were unrelated to chemical reactivity. Sanderson [11] pointed out that the relative average electronic densities of the atoms of the elements can be used as a tool for the measurement of the electronegativity. The SR thus can be estimated from the knowledge of electronegativity. The relation of SRs to Pauling electronegativities has previously been presented graphically (Sanderson, 1952), but the numerical dissimilarity has made difficult a precise quantitative comparison.

However, an empirical mathematical relationship between the SR and the electronega-tivity in Pauling Scale is suggested as:

$$\chi^{\frac{1}{2}} = 0.21SR + 0.77 \tag{71}$$

where $\chi$ is the electronegativity in Pauling Scale.

It is also opined (Sanderson, 1952) that the SR of an atom is a measure of its elec-tronegativity.

$$SR = \chi_{Sanderson} \tag{72}$$

Sanderson found that halogen atoms bear a linear relationship between electroneg-ativity and experimental electron affinities. He also established a linear relationship between acid/base strength and SR for a particular atom.

Sanderson defined the partial charge of the atom as a function of the difference between the elemental electronegativity and that of the new electronegativity, the equilibrium electronegativity. His original equation dealt with the difference between the SR for the entire molecule and the SR for the atom for which the partial charge is being calculated.

## LEWIS ACID STRENGTH

Smith [52] complied a numerical scale of acid base character for binary oxides by analogy with the Pauling Electronegativity Scale. For such oxides he proposed an em-pirical acid base parameter "a." Bratsch (1984, 1985, [48] point out that this parameter can be used to predict the standard enthalpy of combination of binary oxide to form oxo-salt as:

$$\Delta H^{0}_{comb} = -[a_{A} - a_{B}]^{2} \tag{73}$$

where $a_{A}$ and $b_{B}$ are the Smith acid base parameter.

Bratsch (1984, 1985, [48] suggested that the equalized $\chi_{eq}$ in an oxide may be estimated by $N/\sum[v/\chi]$ where N is the number of atoms in the oxide formula, v is the number of each element in the oxide formula, $\chi$ is the initial pre bound electronegativ-ity of each element on the Pauling scale.

Bratsch (1984, 1985, [48] also gave a linear relationship between Smith oxide acid base scale and Pauling electronegativity as:

$$a = m \chi_{eq} + b \tag{74}$$

The potential change $\delta_{0}$ on combined oxygen was given by Bratsch (1984, 1985, [48] as follows:

$$\delta_{0} = (\chi_{eq} - \chi_{0})/\chi_{0} \tag{75}$$

Thereafter, Bratsch (1984, 1985, [48]) correlated the Smith's acid base parameter with the partial charge as:

$$a = 33.7 \delta_{0} + 9.2 \tag{78}$$

Brown and Skowron [53] proposed a scale to measure the Lewis acid strength ($S_a$) as:

$$S_a = V/N_t \qquad (79)$$

where V is the oxidation state of the cation and $N_t$ is the average of the coordination number to the oxygen observe.

Brown and Skowron [53] suggested that the average Lewis acid strength is the Pauling bond strength averaged over all the compounds in which the cation appears and proposed:

$$Sa = 1.18\chi^2 \qquad (80)$$

where $\chi$ is in Rydbergs unit and $S_a$ is in valence unit(vu).

## ELECTRONEGATIVITY AND THE WORK FUNCTION

The work functions of metals and the electrode potentials of elements are also related to electronegativity. The work function of a metal is the minimum work required to remove an electron at 0K. Gordy and Orville Thomas [54] made an attempt to relate the two parameters and presented an empirical relation between work function $\Phi$ and Pauling electronegativity $\chi$ as:

$$\chi(\text{Pauling}) = 0.44\,\Phi - 0.15 \qquad (81)$$

Subsequently, Conway and Bockris [55], Miedema, Boer, and Chatel [56] and Trasatti [57] modified the electronegativity-work function relation of Gordy and Thomas. All used the same straight line relation proposed by Gordy and Orville Thomas [54]. All the above workers used Pauling Scale to correlate electronegativity and work function but Michaelson [58] made an exploratory test of the relation between Mulliken electronegativity [59] and the work function. He proposed a general equation

$$\Phi = \chi_{\text{Mulliken}} - P \qquad (82)$$

Michaelson [58] called the term P as periodicity parameter; a quantity which is a measure of the difference between atomic and solid state periodicity.

## CALCULATION OF OTHER PERIODIC PARAMETERS

The development of modern chemistry began with the finding of periodic changes in properties of elements leading to the formulation of the periodic table as well as periodicity law of elements. Although the periodic table does not directly follow from the quantum mechanics, it has the powerful chemical organizing power. The concept of the electronegativity can be employed for the prediction of the stability of chemical bonds, in the calculation of the surface atoms properties, reaction enthalpies, and so forth side by side based on the periodic law, we can calculate one periodic parameter from another. Atomic radius is a very important theoretical parameter which defines the space occupied by an atom or an ion is its radius, whose value can be estimated by various theoretical methods. Although it does not have any precisely defined physical sense, it allows us to formulate a useful working hypothesis, often employed in solutions of various problems in chemistry, physics, biology, and others. The periodic table tells us that the electronegativity and radius are inversely related to each other (for more discussion and references please see [6, 60]. Ghosh [60] offered a scale of electronegativity based on the absolute radius of the atom where it is stated that the electronegativity and atomic radius (r) are inversely proportional to each other. He suggested a linear relationship between $\chi$ and $1/r$.

Another very old and one of the most useful periodic parameter is the hardness. The notion of hardness was first introduced by Mulliken [61] when he pointed out that the "Hard" and "Soft" behavior of various atoms, molecules and ions can be conceived during acid-base chemical interaction. Soon after Mulliken's classification, the terms hardness and it's inverse, the softness were in the glossary of conceptual chemistry and implicitly signified the deformability of atoms, molecules and ions under small perturbation. The hardness refers to the resistance of the electron cloud of the atomic and molecular systems under small perturbation of electrical field. An atom or molecule having least tendency of deformation are hard and having small tendency of deformation are soft. In other words, list polarizable means most hard and in such systems the electron clouds are tightly bound to the atoms or molecules. On the contrary most polarizable means least hard and in such systems the electron cloud is loosely bound to the atoms or molecules. Electronegativity though defined in many different ways, the most logical and rational definition of it is the electron holding power of the atoms or molecules (for more discussion and references please see [39–43, 62]). The more electronegative species hold electrons more tightly than the less electronegative species hold. Thus, if we invoke the qualitative definition of hardness stated above and compare with the qualitative definition of electronegativity, the commonality of their conceptual structures and philosophical basis are self-evident.

Ghosh and Islam [39–43] pointed out the commonality in the fundamental nature of hardness and electronegativity—the holding power of the electron cloud by the chemical species. Thus the qualitative views of the origins of hardness and electronegativity nicely converge to the one and single basic principle that they originate from the same source—the electron attracting power of the screened nuclear charge. As it is a fact that the origin and the operational significance of the electronegativity and hardness are the same, we may conjecture that the two periodic parameters are directly related to each other.

We mention here the views of some scientists in this topic.

Putz [63], after rigorous research on electronegativity and hardness, opined out that the hardness and electronegativity are proportional, $\chi \propto \eta$.

Ayers [64] pointed out that the two fundamental atomic parameters, hardness and electronegativity, have the similar expression.

Li, Wang, Zhang, and Xue [65] suggested that the hardness of atoms can be defined as the electron holding energy of atoms per unit volume that is, $\eta_a = \chi_a / r^3$.

One of the values characterizing atoms and free ions is their ionization energy, a value that determines the energy with which an electron interacts with the atomic core containing a nucleus and the remaining electrons. Another similar value is the electron binding energy in the outermost filled electronic shell in atoms of elements in their thermodynamically stable forms. This energy characterizes the atom core, the "ion core" with valence electrons removed, in a state in which the atoms form metallic bonding in metals or atomic bonding. Progresses in spectroscopic methods the above periodic parameters have an important feature of being experimental values. It is not derived from any models and due to advances in spectroscopic methods is determined accurately.

Mulliken [59] Electronegativity Scale correlated the three periodic parameters together as:

$$\chi = (I + A)/2 \tag{83}$$

Thus, the any one among the three periodic parameters can be calculated by the knowledge of other two periodic parameters.

The valence electron density is a parameter which can define and measure the electronegativity of an atom. Nagle [66] found that the cube root of this ratio of the number of valence electrons divided by the polarizability, $(n/\alpha)^{1/3}$, can be used as a measure of electronegativity for all s- and p-block elements (except the noble gases). Ghosh and Gupta [67] also suggested that both polarizability and electronegativity are periodic properties and they are connected by an inverse relationship, $\chi \propto (1/\alpha)^{1/3}$.

## ELECTRONEGATIVITY AND THE HSAB PRINCIPLE

One of the very purposes of the modern conceptual chemistry stands the capacity of modeling and controlling the chemical reaction via theoretical methods. There is also recognized that only with admission of the electronegativity and hardness concepts in the chemical reactivity principle, like Hard and Soft Acid and Base principle [68] has the benefit to describe the several fundamental phenomena on a sound mathematical way. This principle was empirical till the publication of the landmark paper by Parr and Pearson [69]. In this work, based on the assumption that the energy is quadratic and utilizing the concept of absolute hardness, Parr and Pearson attempted to present a theoretical deduction of the HSAB principle. They assumed the formation of A: B from A and B: may be regarded as comprising two components: (i) shift of some charge, $\Delta N$ from B to A and (ii) formation of the actual chemical bond.

Focus primarily on the first effect, they wrote the energy expression for A and B in the molecule as follows:

$$E_A = E_A^0 + \mu_A^0(N_A - N_A^0) + \eta_A(N_A - N_A^0)^2 \tag{84}$$

and

$$E_B = E_B^0 + \mu_B^0(N_B - N_B^0) + \eta_B(N_B - N_B^0)^2 \tag{85}$$

The electron numbers, $N_A$ and $N_B$ are

$$N_A = N_A^0 + \Delta N \tag{86}$$

and

$$N_B = N_B^0 - \Delta N \tag{87}$$

After the formation of the molecule AB, the chemical potentials of A and B are equal in the molecule (electronegativity/chemical potential equalization principle). Thus

$$\mu_A = \mu_A^\circ + 2\eta_A \Delta N = \mu_B = \mu_B^\circ - 2\eta_B \Delta N \tag{88}$$

The shift of charge/electron transfer,

$$\Delta N = (\mu_B^\circ - \mu_A^\circ)/2 \, (\eta_A + \eta_B) \tag{89}$$

as,        $\chi = -\mu$

Thus,

$$\Delta N = (\chi_A^\circ - \chi_B^\circ)/2(\eta_A + \eta_B) \tag{90}$$

The corresponding energy change was calculated as follows:

$$\Delta E = (E_A - E_A^\circ) + (E_B - E_B^\circ)$$

$$= -(1/2)(\mu_B^\circ - \mu_A^\circ)\Delta N \tag{91}$$

or,

$$\Delta E = -(\chi_A^\circ - \chi_B^\circ)^2/4(\eta_A + \eta_B) \tag{92}$$

As the acid must be more electronegative than base, $(\chi_A^\circ - \chi_B^\circ)$ is always positive, an energy lowering results from the electron transfer process. The difference in absolute electronegativity drives the electron transfer, and the sum of the hardness parameter acts as a drag or resistance. In other words, the differences in electronegativity drive the electron transfer and the sum of the absolute hardness parameters inhibits electron transfer.

If both acid and base are soft, $(\eta_A + \eta_B)$ is a small number, and for a reasonable difference in electronegativities, $\Delta E$ is substantial and stabilizing. This explains the HSAB principle, meanwhile, it seems safe to say that it explains a part: soft prefers soft. But if both acid and base are hard, there is little electron transfer and energy stabilization from electron transfer, for a given difference in electronegativities. Parr and Pearson [69] commented—"*This result seems paradoxical*" and there is the need of the second effect—the formation of the chemical bond. Parr and Pearson [69] also commented that the consideration "*soft-soft interactions are largely covalent, and that hard-hard interactions are largely ionic*" is not always novel. Providing $\eta_A$ and $\eta_B$ are both small, the stabilization of A: B adduct is explained [70] by the concept of double bonding. The concept of double bonding resembles with the π-bonding theory of Ahrland, Chatt, and Davies [71–73] who used it for explaining various metal ion-ligand preferences. In case of the adduct formation between a hard acid and a hard base normally "little two-way electron transfer" occur. It should be notated that Pearson also showed that there will be little one-way transfer from B to A, if $\eta_A$ and $\eta_B$ are large. For cationic acids the probability of double bonding is greatly reduced. The main source of bonding will come from ionic bonding or ion-dipole bonding. Neutral molecules are the most likely to have two-way electron transfer. The unbiased values of $(\chi_A^\circ - \chi_B^\circ)$ for the neutral molecule determine the direction of net electron transfer. The total amount of electron transfer is governed by $(\eta_A + \eta_B)$ and a small value of the summation is favorable for maximum covalent bonding.

The concept of electronegativity provides a measure of the intrinsic strength of an acid or base [64, 74, 75]. A strong Lewis acid is a good electron acceptor and has high electronegativity/low chemical potential. A weak Lewis acid has a lower electronegativity than a strong Lewis acid, but a higher electronegativity than a Lewis base. A strong Lewis base readily donates electrons and has a lower electronegativity than a weak Lewis base. These relations are summarized by Ayers [64] as follows:

$$\chi(\text{strong acid}) > \chi(\text{weak acid}) > \chi(\text{weak base}) > \chi(\text{strong base}) > 0 \qquad (93)$$

The perfect electron donor has $\chi = 0$. One can reify the electron-accepting abilities of real molecules by imagining how they would react with a perfect electron donor. Thus, the electronegativity concept plays a dominating role in the principle of Hard and Soft Acid and Base.

One of the most important questions connected with the problem of reactivity of molecules in different environmental conditions is the prediction and interpretation of the preferred direction of a reaction and the product formation. Sekhon [25] examined metathesis reaction of the type $AB + CD = AD + BC$ in terms of the equalized electronegativity values of various species involved in the double exchange reaction. The conclusion was made by him from the study that an exchange reaction proceeds from left to right if the total sum of the equalized electronegativity value of the products is greater than that of reactants.

## THE CONCEPT OF GROUP ELECTRONEGATIVITY

One important application of the electronegativity concept is in the estimation of the electron-withdrawing ability of chemical groups. The idea of group electronegativity is important because the electronegativity concept evolved largely from the desire of organic chemists to understand reaction mechanism in terms of the inductive effects of various functional groups. For several years' synthesis chemists have expressed a desire for a group electronegativity scale. This application requires the ability to account for charges on group.

Garner-O' Neale, Bonamy, Meek, and Patrick [76] extended the original Pauling concept of electronegativity for defining group electronegativity as the power of a group in a molecule to attract electron to it. They pointed out that the groups have a better ability to donate or accept charges than atom and therefore be considered as reservoir of enhanced charged capacity. Hence, a group of atoms as a unit is potentially able to donate or withdraw considerable amounts of charge with a very little effect on itself.

The ability to dissipate charge over several atoms increases as the number of atoms which constitute the group increases. Upon bond formation between two atoms, charge is transferred from one atom to the other. In the case of a chemical group charge is transferred to the central atom because of its bond. The electronegativity of the group will then be the orbital electronegativity of the central atom suitably modified to account for its charge [77].

## SOME OTHER APPLICATIONS OF ELECTRONEGATIVITY

This section deals with some other methods of assessing electronegativity that have been proposed but which, in the view of the authors, are rather less acceptable than those described so far.

The electronegativity concept when correlated with other atomic parameters, it become useful to explain various complicated phenomena such as crystal structure, electron structure, non linear optical polarizability, ultraviolet reflection coefficient and the valence bond photo emission [78].

The concept of electronegativity can be used in the correlation of the chemical shifts in nuclear magnetic resonance (NMR) spectroscopy or isomer shifts in Mössbauer spectroscopy. More convincing are the correlations between electronegativity and chemical shifts in NMR spectroscopy [79] or isomer shifts in Mössbauer spectroscopy [80]. Both these measurements depend on the s-electron density at the nucleus, and so are a good indication that the different measures of electronegativity really are describing "the ability of an atom in a molecule to attract electrons to itself."

A systematic correlation between electronegativity and values of Tc for superconductive elements, binary alloys and high Tc-oxides was reported [81–83]. Superconductive elements have electronegativity values that are concentrated within a range near the center of Pauling's Scale (1.3–1.9).On the same scale they lie between 2.5 and 2.65 for the high Tc-oxides. Ichikawa found a trend where the Tc values for some high Tc-oxides increase systematically with weighed harmonic mean electronegativity ($c_{WH}$) in such a way that they head towards some optimum value which lies between 4.69 and 4.88 eV. Furthermore, a wide range of phenomena (for more discussion and references please see [6, 60]) such as ligand field stabilization, the first filling of p orbitals, the transition-metal contraction, and lanthanide contraction can be understood in terms of electronegativity. The refractive index of silica polymorphs is related with the average electronegativity. A large average c, that is, compact electron cloud will give rise to small polarizability and hence small refractive index. The concept of electronegativity is exclusively used in the coordination chemistry. The electronegativity is a measure of the chemical reactivity of an atom, ion, radical, or molecule, which gives the direction of the electron flow an estimate of the initial amount of the amount of the "charge" transferred which is closely related to the energy barriers for reaction and to the strength of the coordinate bond formation.

The scale and concept of electronegativity is applied for exploring the structure-property relationship of materials. The construction of the structure activity property relationship based on the electronegativity to predict the properties of materials and further design new materials is a major trend in the development of the electronegativity. Rare earth luminescent materials have been intenshively investigated due to their applications in the field of lasers and medicine [78]. The concept of electronegativity is also very important in the research of rare earth material because both the valence charge and the charge transfer energy are related to the electronegativity [78].

Devautour et al. [82] proposed a new interpretative method for thermally stimulated depolarization current (TSDC) measurements. By applying the concept of electronegativity equalization to TSDC results, it is showed [82] that it is possible to obtain an experimental evaluation of the chemical potential of the electrons of the exchanged cation and of the host site in zeolites. This step leads to an evaluation of the fundamental parameters, such as the effective hardness and electronegativity of the sites of zeolites. Such an approach gives an evaluation of the heterogeneity of the aluminosilicate surface and is applied to an exchanged hydrogen mordenite containing various amounts of substituted lithium ions or sodium ions.

Schaeffera et al. [83] have applied the empirical relationship between electronegativity and effective work function to a diverse set of multi-element electrode materials on hafnium dioxide ($HfO_2$) gate dielectrics. To accommodate the multi-element nature of metal gate electrodes, the group electronegativity of the metal was calculated from the geometric mean of electronegativity with respect to the volume stoichiometry of the constituent elements. Their finding suggested that the group electronegativity concept is also extended to work function engineering via dielectric capping materials. The electronegativity trends provide insight into the relative charge neutrality levels of candidate dielectric capping materials and their subsequent impact on the metal effective work function.

Baeten and Geerlings [84] used the electronegativity equalization principle to study the charge distributions in enzymes.

Ramsden [85] studied the influence of electronegativity on the triangular three-centre two-electron bonds. Recently, Reddy et al. [86] showed the correlation between the optical electronegativity and the refractive index of ternary chalcopyrites, semiconductors, insulators, oxides, and alkali halides.

Douillard et al. [87] obtained the solid surface tension of ideal crystals of talc and chlorite. From this result, it is possible, using thermodynamic models, to calculate the heat of immersion in water of these solids and to compare with experimental data obtained for well-known samples. Their study confirmed that the differences between surfaces of talc and chlorite and confirming that a route of calculation of surface tension using electronegativity equalization is very simple and correct.

Kwon et al. [88] opined that the electronegativity and chemical hardness are the two helpful concepts for understanding oxide nanochemistry.

Makino [89] pointed out that the band gap, heat of formation and structural mapping for sp-bonded binary compounds can be interpreted on the basis of bond orbital model and orbital electronegativity.

A relationship between dehydroxylation temperature and electronegativity was suggested by Ray et al. [90].

The relationship between the electronegativity and the charge-injection barrier at organic/metal interfaces was suggested by Tang, Lee, Lee, and Xu [91].

Portier, Campet, Etourneau, Shastry, and Tanguy [92] studied the exclusive role of electronegativity in the materials design. Nowaday, the electronegativity equalization methodology, EEM, [93–96] is frequently used to calculate the charge distribution and reactivity index, for example, local softness and hardness [97, 98], condensed Fukui function [99], electrophilicity index [99, 100] of molecules. Chemical Reactivity Theory (CRT) contains reactivity indices defined as first and second derivatives of ground-state properties with respect to electron number such as the electronegativity and the hardness [63, 101].

## CONCLUSION

The electronegativity is very insightful concept in the theory of chemical bonding-explaining the formation, stability and the structure of molecules and it is quite natural that it assumes a finer structure as the concept of valence bonding develops and evolves with time. Some very fundamental quantities of both inorganic, organic, and physical chemistry are the concept of bond energies, polarities, and the inductive effects, and so forth can only be conceived in terms of electronegativity. The work demonstrates the fact that in spite of the empirical nature of the electronegativity, it is one of the most useful theoretical descriptor of science. We have made an attempt to explain the theoretical basis for the electronegativity equalization principle to address the reactivity of molecular complexes. We have discussed the major applications of the electronegativity in the real world and the relationship with other periodic parameters. The present study formulates the concept of electronegativity as an unchanging fundamental immanent characteristic of a chemical entity. However, the attempts to refine electronegativity theory are not yet sufficiently complete to enable a verdict to be reached on their efficacy; one might reasonably expect that a clearer understanding of electronegativity property will be gained with their further development.

## ACKNOWLEDGMENTS

We wish to express our sincere thanks to Professor D.C. Ghosh, Dr. Putz, and Dr. Ayers for their invaluable discussions and comments on this subject.

## KEYWORDS

- **Dehydroxylation**
- **Electronegativity**
- **Hybrid orbit**
- **Polarizability**

# Chapter 3

## Starch Nanocomposite and Nanoparticles: Biomedical Applications

Mohammad Reza Saboktakin

### INTRODUCTION

Nanotechnology focuses on the characterization, fabrication, and manipulation of biological and non biological structures smaller than 100 nm. Structures on this scale have been shown to have unique and novel functional properties. The potential benefits of nanotechnology have been recognized by many industries, and commercial products are already being manufactured, such as in the microelectronics, aerospace, pharmaceutical, food, and cosmetic industries. Nanoparticles (NPs) have been prepared mainly by methods like salting out, spontaneous emulsification/diffusion, solvent evaporation, polymerization, and nanoprecipitation. In addition, electrospraying or the supercritical technology has shown to be capable of producing uniform particles of less than 100 nm. Nanoparticles can be prepared from a variety of materials such as protein, polysaccharide, and synthetic polymers. The selection of matrix depends on the size of nanoparticle required, surface characteristics, degree of biodegradability, biocompatibility, and toxicity. Among biologically inspired nanocomposites, polysaccharides are probably among the most promising sources for the production of nanoparticles [1–18].

Nanotechnology has emerged as one of the most fascinating area in the biopolymer research which find exciting enhancement of applications in drug delivery systems, food technology, and cosmetic field. This following discussion is focused on the polysaccharide nanoscience, emphasizes on basic methodologies for preparing biopolymer-based nanomaterials. Polysaccharide or biopolymer particles may be formed by promoting self association or aggregation of single biopolymers or by inducing phase separation in mixed biopolymer systems. As time goes on, more polysaccharide-based nanoparticles emerge, which greatly enriches the versatility of nanoparticle carriers in terms of category and function [19–27].

Research on the nanoparticles of polysaccharides such as chitosan, alginate, and glucomannan, which are biocompatible, nontoxic, hence used for different applications in medical, food, and cosmetic industries growing very fast. Natural polysaccharides, due to their outstanding merits, have received more and more attention in the field of drug delivery systems. For the application of these naturally occurring polysaccharides for drug carriers, issues of safety, toxicity, and availability are greatly simplified. All these merits endow polysaccharides a promising future as biomaterials. According to structural characteristics, the polysaccharide nanoparticles are prepared mainly by four mechanisms, namely covalent cross-linking, ionic cross-linking, polyelectrolyte complexation, and self-assembly of hydrophobically modified polysaccharide. Polyelectrolyte polysaccharides can form polyelectrolyte complexation with oppositely charged polymers by intermolecular electrostatic interaction. Polysaccharide-based polyelectrolyte complexation nanoparticles can be obtained by means of adjusting the molecular weight of component polymers in a certain range. In theory, any polyelectrolyte could interact with polysaccharides to fabricate polyelectrolyte complexation nanoparticles. But in practice, these polyelectrolytes are restricted to those water-soluble and biocompatible polymers in view of safety purpose. In recent years, numerous studies have been carried out to investigate the synthesis and the application of polysaccharide-based self aggregate nanoparticles as drug delivery systems. When hydrophilic polymeric chains are grafted with hydrophobic segments, amphiphilic copolymers are formed. Upon contact with an aqueous environment, polymeric amphiphiles spontaneously form micelles or micelle like aggregates via intra or intermolecular associations between hydrophobic moieties, primarily to minimize interfacial free energy. These polymeric micelles exhibit unique characteristics, such as small hydrodynamic radius (less than microsize) with coreshell structure, unusual rheology, thermodynamic stability, depending on the hydrophilic/hydrophobic constituents. In particular, polymeric micelles have been recognized as a promising drug carrier, since their hydrophobic domain, surrounded by a hydrophilic outer shell, can serve as a reservoir for various hydrophobic drugs [28–32].

Hydrogels are three-dimensional, hydrophilic, polymeric networks capable of imbibing large amounts of water or biological fluids. These networks can be classified into two main categories according to the type of cross-linking among the macromolecules, whether it is chemically or physically based. Because of their ability to retain a significant amount of water, hydrogels are quite similar to natural living tissues, rendering them useful for a wide variety of biomedical applications. Among the numerous polymers that have been proposed for the preparation of hydrogels, polysaccharides have a number of advantages over the synthetic polymers which were initially employed in the field of pharmaceutics.

A number of pioneering studies have greatly contributed to our present understanding of polysaccharide hydrogel networks. The physically cross-linked gels are of great interest, particularly because the gel formation can be often carried out under mild conditions and in the absence of organic solvents. This peculiarity allows a very wide range of applications, and derivatizations which further increases their versatility. As a consequence, an ever increasing number of publications and patents concerning hydrogels prepared from native and derivatised polysaccharides are there. These hydrogels have wide applications in different areas like encapsulation of living cells, biologically friendly scaffolds in tissue engineering, sustained-release delivery systems, biosensors, and so on. Environmentally, sensitive hydrogels have the ability to sense changes of pH, temperature, or the concentration of metabolite and release their load as a result of such a change. Thus, hydrogels that undergo physicochemical changes in response to applied stimuli, such as biomolecular binding, are promising materials for drug delivery and tissue engineering. Recent research has proved that hydrogels with such additional functionalities offer highly specific bioresponsiveness. Nanosize hydrogels (nanogels), which are composed of nanoscale gel particles, have attracted growing interest with respect to their potential application in drug delivery systems. Ayame et al recently developed a self-assembly method for preparing physically cross-linked nanogels (<50 nm) through the controlled association of hydrophobically modified polymers in water. Microscale hydrogels of controlled sizes and shapes are useful for cell-based screening, *in vitro* diagnostics, tissue engineering, and drug delivery in a sustained manner [33–50].

## STARCH

Starch is an important naturally occurring polymer of glucose, with diverse applications in food and polymer science, found in roots, rhizomes, seeds, stems, tubers, and corms of plants, as microscopic granules having characteristic shapes and sizes. Each starch typically contains several million amylopectin molecules accompanied by a much larger number of smaller amylose molecules. The largest source of starch is corn (maize) with other commonly used sources being cereals (e.g., corn, wheat, rice, oat, barley) contain 60–80%, legumes (e.g., chickpea, bean, pea) 25–50%, tubers (e.g., potato, cassava, cocoyam, arrowroot) 60–90%, and some green or immature fruit (e.g., banana, mango) contain 70% starch in dry base . Genetic modification of starch crops has recently led to the development of starches with improved and targeted functionality. Annual worldwide starch production is 66.5 million tons and the growing demand for starches has created interest in identifying new sources and modifications or derivatives of this polysaccharide.

### Amylose

Starch is tightly and radially packed into dehydrated granules with origin-specific shape and size. Granules contain both crystalline and amorphous areas, consists of two types of molecules, amylose (normally 20–30%) and amylopectin (normally 70–80%). The granules are insoluble in cold water, but grinding or swelling them in warm water causes them to burst. Both amylose and amylopectin consist of polymers of $\alpha$-D-glucose units in the conformation. The relative proportions of amylose to amylopectin and $\alpha$-(1, 6) branch-points both depend on the source of the starch. The starch granule absorbs water; they swell, lose crystallinity and leach out amylose. The higher the amylose content, the lower is the swelling power and the smaller is the gel strength for the same starch concentration. Of the two components of starch, amylose has the most useful functions as a hydrocolloid. Its extended conformation causes the high viscosity of water soluble starch and varies relatively little with temperature. The extended loosely helical chains possess a relatively hydrophobic inner surface that is not able to hold water well and more hydrophobic molecules such as lipids and aroma compounds can easily replace this. Amylose forms useful gels and films. Its association and crystallization (retrogradation) on cooling decreases its storage stability, causing shrinkage and the release of water (syneresis).

According to X-ray studies starch can be classified to A, B, and C forms. In the native granular forms, the A pattern is associated mainly with cereal starches, while the B form is usually obtained from tuber starches.

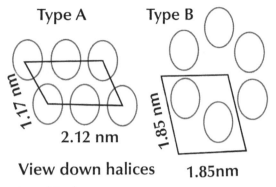

Figure 3.1. Crystalline nature of starch.

The C pattern is a mixture of both A and B types, but also occurs naturally, for example smooth-seeded pea starch and various bean starches. The V-type conformation is a result of amylose being complexed with substances such as aliphatic fatty acids, emulsifiers, butanol, and iodine. The main difference between A and B types is that the former adopt a close-packed arrangement with water molecules between each double helical structure, while the B-type is more open, there being more water molecules, essentially all of which are located in a central cavity surrounded by six double helices (Fig. 3.1).

Amylose molecules consist of single mostly unbranched chains with 500–20,000 $\alpha$-(1, 4)-D-glucose units dependent on source (Scheme 3.1).

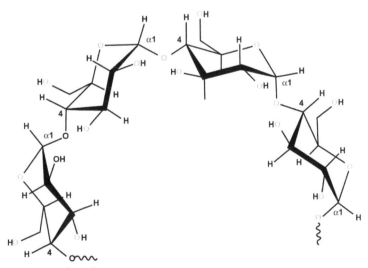

Scheme 3.1. Structure of amylose.

Amylose can form an extended shape (hydrodynamic radius 7–22 nm) but generally tends to wind up into a rather stiff left-handed single helix or form even stiffer parallel left-handed double helical junction zones. Single helical amylase has hydrogen-bonding $O_2$ and $O_6$ atoms on outside surface of the helix with only the ring oxygen pointing inwards. Hydrogen bonding between aligned chains causes retro-gradation and releases some of the bound water (syneresis). The aligned chains may then form double stranded crystallites that are resistant to amylases. These possess extensive inter- and intra-strand hydrogen bonding, resulting in a fairly hydropho-bic structure of low solubility. Single helix amylose behaves similar to the cyclodex-trins, by possessing a relatively hydrophobic inner surface that holds a spiral of water molecules, which are relatively easily to be replaced by hydrophobic lipid or aroma molecules. It is also responsible for the characteristic binding of amylose to chains of charged iodine molecules.

## Amylopectin

Amylopectin is formed by non-random α-(1, 6) branching of the amylose-type α-(1, 4)-D-glucose structure (Scheme 3.2). Each amylopectin molecule contains a million or so residues, about 5% of which form the branch points. There are usually slightly more outer unbranched chains (called A-chains) than inner branched chains (called B chains). There is only one chain (called the C-chain) containing the single reducing group. The-chains generally consist of residues between 13 and 23. There are two main fractions of long and short internal B-chains with the longer chains (greater than about 23–35 residues) connecting between clusters and the shorter chains similar in length to the terminal A-chains. Each amylopectin molecule contains up to two million glucose residues in a compact structure with hydrodynamic radius of 21–75 nm. The molecules are oriented radially in the starch granule and as the radius increases so does the number of branches required in filling up the space, and the consequent formation of concentric regions of alternating amorphous and crystalline structures.

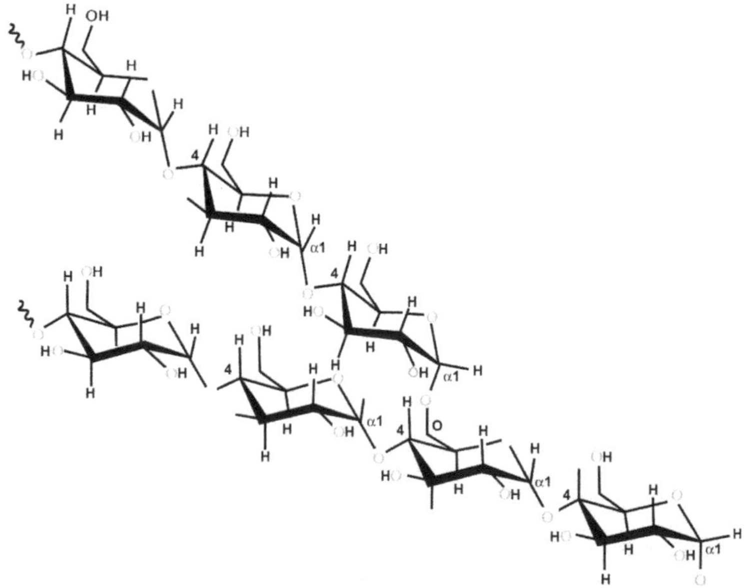

Scheme 3.2. Structure of amylopectin.

Amylopectin double-helical chains can either form the more open hydrated type B hexagonal crystallites or the denser type A crystallites, with staggered monoclinic packing, dependent on the plant source of the granules.

## Modification of Starch

Current starch research has focused on the search for non-conventional starch sources with diverse physicochemical, structural, and functional characteristics that provide them with a broad range of potential industrial uses. Physicochemical and functional properties must be identified before determining the potential uses of starches in food systems and other industrial applications. A fundamental characteristic of native starches from different vegetable sources is that their granule size distribution and molecular structures. These properties then influence a starch's usefulness in different applications.

Native and modified starches are used widely in food processing operations in order to impart viscosity and texture. In their native form, gels or pastes of starches tend to breakdown either from prolonged heating, high shear, or acidic conditions and also they have the tendency to retrograde and undergo syneresis. Starch derivatisations like etherification, esterification, acetylation, and cross-linking have been used to improve the gelatinization and cooking characteristics and to prevent retrogradation. Each anhydroglucose unit of starch contains two secondary hydroxyls and a primary hydroxyl group. These hydroxyls potentially are able to react with any chemical capable of reacting with alcoholic hydroxyls. This would include a wide range of compounds such as acid anhydrides, organic chloro compounds, aldehydes, epoxy, ethylenic compounds, and so forth where the specific chemical contains two or more moieties capable of reacting with hydroxyl groups. There is a possibility of reacting at two different hydroxyls resulting in cross linking between hydroxyls on the same molecule or on different molecules. The most common modification to starches to impart structural integrity is chemical cross-linking. These chemically cross-linked starches are usually resistant to shear, pH, temperature during food processing conditions. The most widely used cross-linking reagents for modifying food starches are mixtures of adipic/ acetic anhydride, and phosphorus oxychloride, or sodium trimetaphosphate, which yield distarch adipates, distarch phosphates, respectively. Cross linking reinforces the hydrogen bonds in the granule with chemical bonds which act as bridge between the molecules. Several research groups have studied the rheological characteristics of starches mixed with other hydrocolloids to improve their rheological properties. Commonly used hydrocolloids are gum arabic, guar, carboxymethylcellulose (CMC), carrageenan, xanthan, xyloglucan, and so forth.

The literature regarding the rheology of starch-hydrocolloid systems has attracted more attention than other systems. Blends of native starches and other polysaccharide hydrocolloids have been used in the modern food industry to modify and control the texture, improve moisture retension, control water mobility, and eating quality of food products. Polysaccharide blending affected the pasting, and rheological properties of cationic, native, and anionic starches differently. Components compatibility, coupled with their individual properties like gelation, ageing, and so forth leads to a large variety of structures and properties of prepared materials. The interactions between starch and hydrocolloid make possible enhancing of viscosity and it has been explained in different ways. Some authors have simplified the system neglecting the granular phase and have explained the enhancement of blend viscosity on the basis of complexation between soluble starch and the added hydrocolloid, while others based on the two-phase model, have suggested that the increase observed in starch–hydrocolloid mixtures can be attributed to an artificial increase of hydrocolloid concentration in the continuous phase due to the reduction of its volume by swelling of the starch granules during gelatinization. Starch and starch derivative films have been widely studied from 1950 due to their great molding and film forming properties, high oxygen barrier, and good mechanical. Even though there have been numerous studies conducted on the properties of starch-based films, few studies have related starches from different sources with the resulting film forming characteristics, mechanical, and physical properties. Among the starch films, potato, sweet potato, mungbean, and waterchestnut were selected due to their superior film forming properties when compared with synthetic films. Forssell et al. reported that starch-based polymer films, plasticized with water only had good oxygen barrier properties under ambient humidity. Since water acts as a plasticizer for these materials and their mechanical and barrier properties strongly depend on water content. The HACS forms strong and flexible films probably due to amylose crystallization. Amylose is responsible for the film-forming capacity of starch-based films. The addition of plasticizing agent to edible films is required to overcome film brittleness caused by extensive intermolecular forces, thereby improving flexibility and extensibility of films. Plasticizers extend, dilute, and soften the structure and increasing the chain mobility [51–69].

## Starch Micro/Nanoparticles

Starch microspheres have been found to be effective in the systemic delivery of peptides after nasal administration and of vaccine given orally and intramuscularly. Pharmaceutical applications of starch microspheres necessitate controlled particle size, generally narrow size distribution, because the localization and distribution of the particles in the body depend on these parameters. Most studies based on the use of starch microspheres have been made with Spherex microspheres which have been commercialized since 1994 by pharmacia. These particles were prepared by the action of epichlorohydrin on partially hydrolysed starch. Hamidi et al. prepared starch-based microparticles by a water-in-water (w/w) emulsification-cross-linking method for different application. The main part of the starch particles has been produced by polymerization of acryloylated starch in water in oil emulsion or by cross-linking soluble starch with epichlorohydrin. The emulsion method is considered as best method for control the size of starch particles. In these cases, the amount of surfactant, epichlorohydrin/starch molar ratio and so forth are very important. Starch acetate microparticles are also used widely for targeted drug delivery application [70–87].

In spite of great potential as a bioadhesive carrier, starch microspheres are not normally sufficient to provide clinically relevant plasma levels of large polypeptides. To address above limitations, extensive studies were carried out on microsphere formulations comprising permeation enhancers and few studies dealt with nanoparticles too. Using a new approach developed at ATO-DLO, it was shown that a novel type of starch-based micro or nanoparticles could be prepared which behaved as colloids in aqueous solution. The synthesis of the particles was based on a unique combination of gelatinization and cross-linking, performed in water-oil emulsions. Starch-based nanoparticles with variations in sizes, charge, density, and suspension properties were prepared by varying starch source, cross-linker, pH, emulsion type, energy, temperature, and other parameters. These materials are biodegradable and expected to be applicable in both the food area and the non-food area.

Chakraborty et al. studied the solution properties of starch nanoparticle in water and DMSO using DLS and selective esterification of starch nanoparticles was performed using as catalyst *Candida antartica* Lipase B in its immobilized and free forms. The starch nanoparticles were made accessible for acylation reactions by formation of AOT-stabilized microemulsions. Starch nanoparticles in microemulsions were reacted with vinyl stearate, ε-caprolactone (CL), and maleic anhydride at 40°C for 48 h to give esterified starch nanoparticles [88–93].

*Nanoparticles made from novel starch derivatives*
In the last years, the use of polysaccharides to prepare nanoparticles has increased considerably. Polysaccharides possess many recognition functions, allowing, for example, mucoadhesion or specific receptor recognition, as well as providing neutral coatings with low surface energy, preventing non specific protein adsorption. On the other hand, the high amount of hydroxyl groups in the polysaccharide backbone allows the incorporation of different specific ligands to obtain polyfunctional colloidal systems. In this context, starch has an interesting potential, which is so far relatively unexplored. Starch is a biocompatible, biodegradable, nontoxic polymer, abundantly occurring in nature as the major polysaccharide storage in higher plants. However, despite these properties, some problems can occur. The hydrophilic nature of starches is a major constraint that seriously limits the development of starch-based nanoparticles. A good alternative to solve this problem is the grafting of hydrophobic side chains to the hydrophilic starch backbone. However, an important restriction usually arising from the use of these hydrophobic polysaccharides derivatives to prepare nanoparticles, is the necessity to employ organic solvents, such as dichloromethane or dimethyl sulfoxide, with considerable toxicological and other safety risks.

Besides, an increasing number of researchers are investigating on the improvement of percutaneous drug absorption. However, the excellent barrier properties of the skin is a handicap that must be overcome. Usually, to reach a high and constant drug flux through the skin a change of the skin barrier function is necessary. However, in the last years, different authors have demonstrated that the use of particles as transdermal drug delivery systems (TDDS) enhance the rate and extent of transport across skin, without compromising the skin barrier function. Although such systems were undoubtedly able to enhance skin penetration and distribution, the mechanism by which this enhancement was achieved is still unclear.

The aim of this work was to explore nanoparticulate drug carriers based on starch derivatives, by using the advantages of hydrophobic starch derivatives. With this in mind, we have selected propyl-starch derivatives instead of the well known acetyl-starch derivatives to formulate our nanoparticles. The inclusion of propyl groups, even with low degree of substitution, would enable good solubility in low hazardous organic solvents, such as ethyl acetate. Moreover, propyl-starches may allow a better quality control since degree of substitution can be determined after acidic degradation to the respective glucose derivatives by 1H-NMR spectroscopy. Two propyl-starch derivatives and one un-modified starch polymer were used as basic constituents for the preparation of nanoparticles, representing two different degrees of substitutions (Ds) 1.05 (PS-1) and 1.45 (PS-1.45), respectively. Nanoparticles were formulated by emulsification–diffusion technique. This method has several advantages such as high yields generally obtained, high batch to batch reproducibility and easy scaling up. Moreover, it is possible to control the size and polydispersity of nanoparticles by the control of the oil/water phase ratio. Once nanoparticles were formed, a thorough physicochemical characterization was carried out. Afterwards, the capacity of these nanoparticles as drug delivery systems was explored by the encapsulation and release of three different model drugs, flufenamic acid (FFA), testosterone (Test), and caffeine (Caff). Finally, considering the nanoparticles characterization results, it was possible to establish the use of these nanoparticles as TDDS. Hence, the potential application of these nanoparticles as TDDS was analyzed by studying their influence on the permeability of human heat-separated epidermis (HSE).

*Enzyme-catalyzed regioselective modification of starch nanoparticles*
Starch is an abundant, inexpensive, naturally occurring polysaccharide. It is biocompatible, biodegradable, and nontoxic, so it can be used as biocompatible implant materials and drug carriers. Literature reports describe the use of chemically modified forms of starch for sustained drug delivery systems. For example, epichlorohydrin cross-linked high amylose starch was used as a matrix for the controlled release of contramid. A complex of amylose, butan-1-ol, and an aqueous dispersion of ethylcellulose was used to coat pellets containing salicylic acid to treat colon disorders. Starch has also been used as a carrier for phenethylamines, acetylsalicylic acid, and estrone. Hydrogels composed of starch/cellulose acetate blends were reported as possible bone cements. While starch-based biomaterials appear promising, scientific challenges remain to be solved. For example, it would be advantageous if starch esters used as matrices for drug delivery could be prepared so that they are modified at selected positions of the glucose residues (i.e., at only the primary or secondary positions). This is difficult due to the presence of three hydroxyl groups per glucose residue each in different chemical environments. Furthermore, to solubilize starch for homogeneous modification, polar aprotic solvents such as dimethyl sulfoxide are needed. For example, to modify the primary (6-O) hydroxyl sites of amylose, starch was heterogeneously persilylated, the persilylated derivative in carbon tetrachloride was acylated with an anhydride, and then the silyl protecting groups were removed. Previous work has investigated the use of enzymes to regioselectively modify polysaccharides under mild conditions. Hydroxyethylcellulose (HEC) particles were suspended in dimethylacetamide and acylated with vinyl stearate using *Candida antarctica* Lipase B (CALB) as catalyst. After 48 h, a product with degree of substitution (DS) 0.1 was formed. Lipase-catalyzed modification of HEC in film or powder form by reaction with CL gave low-DS HEC-*g*-PCL copolymers [94–101].

Problems associated with these strategies are use of (i) polar aprotic solvents that strip critical water from enzymes lowering their activities10 and (ii) heterogeneous reaction conditions that restrict the modification of large particles and films to a small fraction of the substrate residing at the surface. To overcome the use of polar aprotic solvents, enzymes have been incorporated within reverse micelles using the anionic surfactant aerosol-OT [AOT, bis(2-ethylhexyl)sodium sulfosuccinate]. The AOT forms thermodynamic water droplets surrounded by a surfactant monolayer in oil (isooctane).

Water entrapped within the reverse micelles resembles the polar pockets in cells. Incorporation of enzymes within reverse micelles soluble in nonpolar media facilitates productive collisions and reactions between enzymes and nonpolar substrates. Several types of lipase-catalyzed reactions in AOT/isooctane have been studied. Dordick and co-workers incorporated proteases from *Subtilisin carlsberg* and *Bacillus lichenifor-mis* within AOT-coated reverse micelles. Although the enzymes within reverse micelles were active for the acylation of amylose in film and powder form, the inability of AOT-coated enzyme to diffuse into the bulk of these substrates limited modification of films and powders to surface regions. Nanoparticles, nanospheres, and nanogels are used as building blocks for nanoscale construction of sensors, tissues, mechanical devices, and drug delivery systems. For the latter, carriers with nanodimensions are not detected by the reticuloendothelial system so they circulate for longer times. For medical applications, nanoparticles constructed from poly(lactic acid), poly-(glycolic acid), poly(alkyl cyanoacrylate), 2-hydroxyethyl acrylate-poly(ethylene glycol)diacrylate copolymers, poly(L-lysine)-g-polysaccharides, and poly- (vinylpyrrolidone) have been reported. Starch microspheres were studied for the delivery of insulin via the nasal system. To overcome previous difficulties in the enzymatic esterification of polysaccharides and to prepare a family of structurally and dimensionally well-defined nanoparticles from a natural material, a new idea was explored. That is, starch nanoparticles were incorporated into reverse micelles stabilized by AOT. A key discovery was that the nanodimensions of AOT-coated starch particles and their solubility in nonpolar media such as toluene allowed their diffusion to sites within the pores of the physically immobilized lipase catalyst Novozym 435 at which esterification of starch occurred [102–110].

The acyl donors included vinyl esters of fatty acids differing in chain length, maleic anhydride, and $\epsilon$-caprolactone. Motivations for selecting these acyl donors were as follows: (i) increase starch hydrophobicity, (ii) introduce both carboxylate side chains and sites for free radical cross-linking, and (iii) form polyester grafts. Nuclear magnetic resonance experiments gave the DS and regioselectivity of esterification reactions. The influence of the reaction temperature, time, and structure of the acyl donor on the progress of esterification reactions was studied. IR microspectroscopy revealed the extent that AOT-coated starch nanoparticles diffuse through the macroporous structure of the physically immobilized lipase catalyst Novozym 435. Dynamic light scattering (DLS) showed the size distribution of AOT-coated reverse emulsions of starch nanospheres both prior to and after starch modification reactions. Furthermore, DLS confirmed that the nanodimensions of modified starch nanoparticles were retained after they had been stripped of AOT and dispersed in water or DMSO.

*Preparation of folate-conjugated starch nanoparticles*

The lower toxicity and high effect of drug are very important for clinic therapy. So, more and more attention has been paid to the targeted drug delivery system. Folate receptor (FR) has been reported to be vastly overexpressed in most human tumors but seldom ex-pressed on normal tissues, so the folate/FR mediated targeted drug delivery system has been popularly re-searched. Folate, a vitamin with small molecular weight, is stable, and not immunogenic and expensive compared to the monomolecular anti-body. Folate displays extremely high affinity to its receptor, which enables them to rapidly bind to the FR and become inter-nalized via an endocytic process. So, folate is a valuable targeting molecular. For the purpose of sustaining release of drug, folate was conjugated with drug carrier, such as lipids, polymers, protein, and nanoparticles, and then the carriers conjugating with folate possess simultaneously the ability of targeting and sustaining release.

As a natural biomaterial, starch is biocompatible, not immunogenic, stable in air, and does not react with most drugs. The reducing of starch is rapid, and the production of this reducing reaction is deoxidizing sugar, which could be digested easily. As a traditional filling agent, starch was lately found to have more properties through physical or chemic denaturalization, thus it could be used diffusely in medicine field as drug carrier material. By far, there is no report about starch conjugating with folate and used in targeted drug delivery system.

*Novel pH-responsive starch-based nanoparticles*

The new pH sensitive starch-based nanoparticles maybe useful for drug delivery and diagnostic applications. Polymethacylic acid-graft-starch (PMAA-g-St) nanoparticles with various molar ratio of starch to MAA were synthesized using a novel dispersion polymerization method. The method enabled simultaneous grafting and nanoparticle formation in a one-pot synthesis procedure. The grafting was confirmed using FTIR and 1H NMR spectroscopy. The particles morphology was examined using TEM. The particles size and surface charge of the nanoparticles as a function of pH in buffer were determined by DLS and electrophoretic mobility measurements. The effect of formulation and processing parameters such as cross-linker, surfactants, and total monomer concentrations on the size and pH-sensitivity of the nanoparticles were studied using DLS. The capability of nanoparticles for drug delivery was evaluated using mathylene blue as model drug. Drug loading and release from the nanoparticles was measured using UV-Vis spectroscopy. The PMAA-g-St nanoparticles with diameters in a range of 120–600 nm were obtained. The particle size could be tailored by changing the composition and process parameters. The PMAA-g-St nanoparticles were more monodispersed compared to PMAA nanoparticles. The nanoparticles were relatively spherical and had a porous surface morphology. The particles were able to effectively load methylene blue, a cationic drug, and exhibited pH-dependent swelling and drug release in a physiological pH range. The particle size and magnitude of phase transition were dependent on starch content and various processing parameters such as surfactant levels, cross-linker amount, and total monomer concentration. Depending on the ratio of starch to MAA and cross-linking ratio the particles exhibited 1.5–10 fold volume change when pH was increased from 4 to 7.4.

*Starch Nanocrystals*

Starch nanocrystals can be obtained by an acid hydrolysis on starch native granules. They consist of crystalline nanoplatelets about 6–8 nm thick with a length of 20–40 nm and a width of 15–30 nm. These starch nanocrystals displayed interesting reinforcing properties when dispersed in different mediums. Interesting reinforcing capability was also obtained for nanocrystals of reinforced starch plasticized by glycerol. Grafting of larger chains on the surfaces of starch nanocrystals enhance the nonpolar nature of original nanoparticles and dispersion in organic media. The mechanical properties of nanocomposite materials processed from the modified nanocrystals and natural rubber were found to be lower than those for unmodified nanoparticles, because the length of the grafted chains is high enough, entanglements are expected to occur with the polymeric matrix.

*Njavara Rice Starch*

Njavara is a rice variety widespread to Kerala, mainly seen in the northern parts (Fig. 3.2). The cultivation of this rice variety is recorded from 2500 years back. Njavara is a unique grain plant in the Oryza group and widely used in the Ayurvedic system of medicine, especially in Panchakarma treatment. Njavara as a special cereal, have the properties to rectify the basic ills affecting our circulatory, respiratory, and the digestive systems. This variety is highly resistant to drought conditions and is generally resistant to diseases. Dehusked Njavara rice has 73% carbohydrates, 9.5% protein, 2.5% fat, 1.4% ash, and 1628 kJ per 100 g of energy. Higher amounts of thiamine (27–32%), riboflavin (4–25%), and niacin (2–36%) and the total dietary fiber content in Njavara was found to be 34–44% higher than compared to the other rice varieties. Significantly higher phosphorus, potassium, magnesium, sodium, and calcium levels were found in Njavara rice. Two types of Njavara are recognized, the black and golden yellow glumed. In the case of black glumed variety, the seed color is red.

**Figure 3.2.** Njavara rice.

*Applications of Starch*

Starch is a versatile and economical, and has many uses as thickener, water binder, emulsion stabilizer, and gelling agent. Starch is often used as an inherent natural ingredient but it is also added for its functionality. Mixing with hydrocolloids and low molecular weight sugars can also reduce retrogradation. At high concentrations, starch gels are both pseudoplastic and thixotropic with greater storage stability. Their water binding ability can provide body and texture to food stuffs and can be used as a fat replacement. Many functional derivatives of starch are marketed including cross-linked, oxidized, acetylated, hydroxypropylated, and partially hydrolyzed material.

Hydrolysis of starch, usually by enzymatic reactions, produces a syrupy liquid consisting largely of glucose. It is widely used to soften texture, add volume, inhibit crystallization, and enhance the flavor of foods. Nowadays starch-based films are used for food packging, a coating for tablets or capsules. Starch nanoparticles and microparticles were used for the release of drugs, cosmetics, and aromas, and so forth.

## STARCH NANOCOMPOSITES

### Thermoplastic Cassava Starch/Sorbitol-modified Montmorillonite Nanocomposites Blended with Low Density Polyethylene

One of today's serious global problems is the management of the steadily increasing amount of solid waste. Tremendous quantities of polymers, mainly polyolefins (e.g., polyethylene, polypropylene) are produced and discarded into the environment, ending up as wastes that do not degrade spontaneously.

Some polymer products have a short useful life, in many cases of less than 2 years. They are consumed and discarded into the environment when their utilization ceases. The petroleum-based polymers were developed for durability and resistance to all forms of degradation. With developing environmental ecological awareness, biodegradable plastics are proposed as one of many strategies to alleviate the environmental impact of petroleum-based plastics and are gaining public interest. They are designed to be easily degraded by the enzymatic action of living microorganisms such as bacteria, yeasts, and fungi. In contrast to synthetic polymers, natural polymers are good base materials for producing inexpensive, rapid degradable plastics. The use of biodegradable materials based on renewable resources can help reduce the percentage of plastics in industrial and household wastes. Therefore, several considerable efforts have been made to accelerate the biodegradability of polymeric materials by replacing some or all of the synthetic polymers with natural polymers in many applications in order to minimize the environmental problems caused by plastic wastes. Starch is one of the main natural polymers used in the production of biodegradable materials because of its renewability, biodegradability, wide availability, and low cost. However, the starch presents some drawbacks, such as the strong hydrophilic behavior and poorer mechanical properties when compared to synthetic polymers. It is predominantly water-soluble and cannot be processed by melt-based routes because it decomposes before melting. To improve the mechanical and barrier properties of the starch composites at the same time, a small amount of inorganic nanofillers is commonly added to a polymer matrix. The clay, montmorillonite (MMT), is one of the attractive nanofillers utilized due to its high aspect ratio of width/thickness, in an order of 10–1000. Indeed, for very low amounts of nanoparticles, the total interface between polymer and silicate layers is much greater than that in conventional composites. In general, the clay needs to be modified in order to enlarge the interlayer distance. Ma, et al. prepared sorbitol-modified montmorillonite, which was added into the thermoplastic starch (TPS) to obtain nanocomposites by dual-melt extrusion process. Sorbitol is an alcohol sugar widely used in the food industry, not only as a sweetener, but also as a humectant, texturizer, and softener. In this study, cassava starch was mixed with MMT to obtain starch nanocomposites in order to improve its mechanical properties and water resistance. However, the starch was first plasticized under heating to obtain TPS, giving rise to a continuous phase in the form of a viscous melt which can be processed by conventional plastic processing technique. In general, plasticizers used include polyols such as glycerol, glycol, xylitol, and sorbitol. Plasticizers containing amide groups such as urea, formaldehyde, and acetamide or a mixture of plasticizers have also been studied. In this research, the plasticizers used for preparing TPS were sorbitol and formamide. Thermoplastic cassava starch/sorbitol-modified MMT nanocomposites with various amounts of MMT were incorporated into the blend of 100 LDPE/80 PE wax in order to enhance the mechanical properties and biodegradability of the blend. The structure and morphology of the samples were investigated by X-ray diffractometer and transmission electron microscope. The impact strength, flexural strength, and biodegradability were also examined.

### Thermal and Thermomechanical Behavior of Polycaprolactone and Starch/ Polycaprolactone Blends for Biomedical Applications

Polycaprolactone (PCL) is among the most attractive and commonly used biodegradable polyesters. It can be used in different biomedical applications, such as in scaffolds in tissue engineering, and for controlled release of drugs. On the other hand, starch is one of the natural biodegradable polymers and is produced at a relatively low price. The development and biodegradable properties of the blends of starch with PCL (SPCL) has been well documented in the literature. Starch helps to lower the cost of the ultimate product as well as to give some biodegradable characteristics to PCL. Recently, SPCL has already been proposed for biomedical applications, including tissue engineering scaffolds, and for different orthopaedic purposes. Besides adequate physical properties, this blend exhibits good biocompatibility and low inflammatory response.

Characterizing the thermal properties of such systems may be useful for the processing of the material and for the prediction of some features during their potential applications as biomaterials. Non-isothermal crystallization behavior is one of the important thermal properties of semi-crystalline polymers to be characterized, since most processing techniques are melt-based and actually occur under non-isothermal conditions, and the resulting physical properties (including mechanical and biodegradable behavior) are strongly dependent on the morphology formed and the extent of crystallization. Examinations concerned with the non-isothermal crystallization features of PCL and its blends with starch have been published by several authors. For example, Skoglund et al. presented overall crystallization characteristics of PCL. Vazquez et al. reported the influence of sisal fiber on the crystallization behavior of PCL. However, there is little information concerning the influence of starch on the crystallization behavior of PCL. In this article, the melting behavior and the nonisothermal crystallization kinetics of typical commercially available PCL and SPCL were studied by DSC.

The morphology development upon cooling from the melt was also studied by polarized optical microscopy.

The mechanical characterization of new polymeric systems is essential to understand their performance under loads and may help to elucidate on the micro-structure of heterogeneous systems, such as semi-crystalline polymers, blends, or copolymers. Especially for implanted materials that will withstand mechanical stresses in clinical use (e.g., in vascular or orthopaedic applications), a proper mechanical characterization is among the most important physical tests that must be carried out. Implantable materials should have a similar mechanical performance of the living tissues that will be in contact with. Most of the biological tissues, possibly excepting dental enamel and echinoderm skeletons, exhibit a time-dependent mechanical behavior due to their viscoelastic nature. Therefore, it is important to evaluate the solid-state rheological properties of materials aimed at being used in biomedical applications. Dynamic mechanical analysis (DMA) is a thermal analysis technique in which the response of the material under a cyclic load or strain excitation is measured as a function of frequency or temperature, being adequate to probe the viscoelastic properties of polymeric systems. It has also been shown that this technique may be useful to extract relevant information in biomaterials. A few authors have shown some DMA data of PCL and SPCL, but they only reported the results at a single frequency; moreover, the data were never integrated in the context of the potential biomedical applications of the materials. In this work DMA was also used to access the thermal properties of the studied materials, especially near glass transition temperature (Tg), and to obtain information about the viscoelastic properties in this temperature region at meaningful frequencies.

### Hydroxyapatite—Starch Nano Biocomposites Synthesis and Characterization

Chemical formula of Hydroxyapatite (HAp) is $Ca10(PO4)6(OH)2$ which is very similar to the materials forming the bones in the human body. The HAp crystals in natural bone are needle-like or rod-like in shape 40–60 nm in length, 10–20 nm in width, and 1–3 nm in thickness. The synthesized HAp with bone-bonding properties is widely used in hard tissue replacement due to their biocompatibility and osteoconductive properties. Many characteristics of HAp, such as surface characteristics and bioactivity can be affected by the shape of HAp crystal. Therefore, the applications of the HAp can be expanded by controlling the crystal shape of the nanometer HAp, such as needle-like, spherical, plate-like shape and so on. At present, many studies have reported to synthesize the nanometer HAp with different shapes. For example, the needle-like HAp has been synthesized by different processing methods including organic gel systems, homogeneous precipitation, or hydrothermal technology. The rod-like HAp has been synthesized by precipitating calcium nitrate tetra hydrate and ammonium dibasic phosphate in the presence of polyacrylic acid followed by hydrothermal treatment. Brittleness of HAp limits its use. One of the methods to solve the problem is combination of it with polymer. Surfaces of organic materials can be tailored to achieve different properties, such as the capability of carrying functional groups, chelate to metal ions by their functional groups and hydrophilicity [9]. For instance, major approach in the development of materials for bone regeneration and replacement is the use of degradable polymers as matrices. Biodegradable materials have to degrade without an unresolved inflammatory response or an extreme immunogenicity or cytotoxicity. Some attempts via in situ mineralization technique have been done using polymeric additives such as poly (vinyl alcohol) poly (lactic acid) (PLA), poly (acrylic acid) (PAAc), collagen, and due to their calcium binding properties. Biopolymers are an important source of materials for biomedical applications. One of the biopolymers with biomedical applications is starch. This natural polymer is biodegradable, biocompatible, water soluble, and inexpensive in comparison with other biodegradable polymers.

### Synthesis and Characterization of Silver/Clay/Starch Bionanocomposites by Green Method

Bionanocomposites (BNCs), a novel invention of nanocomposite materials, indicate a promising field in the frontiers of nanotechnology, materials, and life sciences. The BNCs are composed of a natural polymer matrix and organic/inorganic filler with at least one dimension on the nanometer scale. In addition to these characteristics, BNCs show the extraordinary advantages of biocompatibility and biodegradability in various medical, drug release packaging, and agricultural applications.

Among natural polymers, starch is one of the most promising biocompatible and biodegradable materials because it is a renewable resource that is universally available and of low cost. A number of researchers have presented work in the field of starch-based BNCs, which can be obtained by filling a TPS matrix with nanofillers such as layer silicates; MMT and kaolinite are the usual layer silicates used in starch-based BNCs. The MMT as lamellar clay has intercalation, swelling, and ion exchange properties. Its interlayer space has been used for the synthesis of material and biomaterial nanoparticles, as support for anchoring transition-metal complex catalysts and as adsorbents for cationic ions. Silver nanoparticles (AgNPs) possess many interesting and unique properties. It is found in various applications, such as catalysis, electronics, non-linear optics, antimicrobial, and biomaterial applications. Several methods have been reported for the synthesis of AgNPs, for example, photochemical reduction, microwave, chemical reduction, and γ-irradiation. The concept of green AgNPs preparation was first developed by Raveendran, who used β-D-glucose as the reducing agent and starch as a capping agent to prepare starch AgNPs. A green method for nanoparticles preparation should be evaluated from three aspects: the solvent, the reducing agent, and the stabilizing agent. In this chapter, we present a simple and green method for the synthesis of Ag/MMT/Stc BNCs. In here MMT, Starch, β-D-glucose and AgNO$_3$ were used as a solid support, stabilizer, green reducing agent, and silver precursor, respectively. The TEM images show that AgNPs prepared in the edge and extra surface of MMT layers have large size (39 ± 14.09 nm) and AgNPs intercalated between MMT layers have small size (9 ± 3.39 nm). The Ag/MMT/Stc BNCs are very stable in aqueous solution over a long period of time (i.e., three months) without any sign of precipitation and have potential for various medical applications.

### Starch-based Nanocomposites Reinforced with Flax Cellulose Nanocrystals

The development of commodities derived from petrochemical polymers has brought many benefits to mankind. However, it is becoming more evident that the ecosystem is considerably disturbed and damaged as a result of the non-degradable plastic materials used in disposable items. Therefore, the interest in polymers from renewable resources has recently gained exponential momentum and the use of biodegradable and renewable materials to replace conventional petroleum plastics for disposable applications is becoming popular. Within the broad family of renewable polymers, starch is one of the most attractive and promising sources for biodegradable plastics because of the abundant supply, low cost, renewability, biodegradability, and ease of chemical modifications. In recent years, plasticized starch (PS) has attracted considerable attention and has offered an interesting alternative for synthetic polymers where long-term durability is not needed and rapid degradation is an advantage. However, compared with conventional synthetic thermoplastics, biodegradable products based on starch, unfortunately, still exhibit many disadvantages, such as water sensitivity, brittleness, and poor mechanical properties. Various physical or chemical means have been used to solve these problems, including blending with other synthetic polymers, the chemical modification, graft copolymerization, and incorporating fillers such as lignin, clay, and multi walled carbon nanotubes.

More recently, there is an increased use of cellulose nanocrystals (CNs) as the loading-bearing constituent in developing new and inexpensive biodegradable materials due to a high aspect ratio, a high bending strength of about 10 GPa, and a high Young's modulus of approximately 150 GPa. The CNs from various sources such as cotton, tunicate, algae, bacteria, ramie, and wood for preparation of high performance composite materials have been investigated extensively. Both natural and synthetic polymers were explored as the matrixes. Natural polymers such as poly(β-hydroxyoctanoate) (PHO), soy protein, silk fibroin reinforced with cellulose whiskers were reported. Meanwhile, Poly-(styrene-*co*-butyl acrylate) (poly(S-*co*-BuA)), poly(vinyl chloride) (PVC), polypropylene, waterborne polyurethane, were also used as synthetic matrixes. The flax plant (*Linum usitatissimum*) is a member of the Linaceae family, which is an important crop in many regions of the world. Fibers from flax have been used for thousands of years to make different textile products because of their excellent fiber characteristics. Therefore, the search on short fibers from flax as a replacement for synthetic fibers in many non-textile products, for example, in polymer compounds, building materials, and absorbent materials, has attracted much attention in the last decade.

However, incorporating flax cellulose nanocrystals (FCNs) in the starch-based composite films has not been reported in the literatures. In present work, it attempts to prepare CNs from flax fiber by acid hydrolysis with concentrated sulfuric acid, and then use the resulting CNs to reinforce PS for preparation of nanocomposite films with improved performances.

The resulting films were prepared by casting the mixture of aqueous suspensions of CNs and PS in various weight ratios. The morphology, structure, and performance of the resulting nanocomposite films were investigated by scanning electron microscopy (SEM), wide-angle X-ray diffraction (WAXD), differential scanning calorimetry (DSC), and measurement of the mechanical properties and water uptake.

### Starch-based Completely Biodegradable Polymer Materials

As well known, synthetic polymer materials have been widely used in every field of human activity during last decades, that is post-Staudinger times. These artificial macromolecular substances are usually originating from petroleum and most of the conventional ones are regarded as non-degradable. However, the petroleum resources are limited and the blooming use of non-biodegradable polymers has caused serious environmental problems. In addition, the non-biodegradable polymers are not suitable for temporary use such as sutures. Thus, the polymer materials which are degradable and/or biodegradable have being paid more and more attention since 1970s.

Both synthetic polymers and natural polymers that contain hydrolytically or enzymatically labile bonds or groups are degradable. The advantages of synthetic polymers are obvious, including predictable properties, batch-to-batch uniformity and can be tailored easily. In spite of this, they are quite expensive. This reminds us to focus on natural polymers, which are inherently biodegradable and can be promising candidates to meet different requirements. Among the natural polymers, starch is of interest. It is regenerated from carbon dioxide and water by photosynthesis in plants. Owing to its complete biodegradability, low cost and renewability, starch is considered as a promising candidate for developing sustainable materials. In view of this, starch has been receiving growing attention since 1970s. Many efforts have been exerted to develop starch-based polymers for conserving the petrochemical resources, reducing environmental impact, and searching more applications.

### Preparation of starch-based biodegradable polymers

*Physical blend with synthetic degradable polymers*

To improve the properties of starch, various physical or chemical modifications of starch such as blending, derivation, and graft copolymerization have been investigated. At first, starch was adopted as a filler of polyolefin by Griffin and its concentrations is as low as 6–15%. Attempts to enhance the biodegradability of the vinyl polymers have been investigated by incorporating starch to a carbon–carbon backbone matrix. In all these cases starch granules were used to increase the surface area available for attack by microorganisms. However, such a system is partially biodegradable and not acceptable from an ecological point of view. Thus, the blends of starch and polyolefin will not be mentioned any more. To prepare completely biodegradable starch-based composites by this strategy, biodegradable polymers are assumed. Usually, the components to blend with starch are aliphatic polyesters, polyvinyl alcohol (PVA) and biopolymers. The commonly used polyesters are poly(β-hydroxyalkanoates) (PHA), obtained by microbial synthesis, and polylactide (PLA) or poly(ε-caprolactone) (PCL), derived from chemical polymerization. The goal of blending completely degradable polyester with low cost starch is to improve its cost competitiveness whilst maintaining other properties at an acceptable level.

The PLA is one of the most important biodegradable polyesters with many excellent properties and has been widely applied in many fields, especially for biomedical one. PLA possesses good biocompatibility and processability, as well as high strength and modulus. However, PLA is very brittle under tension and bend loads and develops serious physical aging during application. Moreover, PLA is a much more expensive material than the common industrial polymers.

Many efforts have been made to develop PLA/starch blends to reduce total raw materials cost and enhance their degradability. The major problem of this blend system is the poor interfacial interaction between hydrophilic starch granules and hydrophobic PLA. Mechanical properties of blends of PLA and starch using conventional processes are very poor because of incompatibility. In order to improve the compatibility between hydrophilic starch granules and hydrophobic PLA, glycerol, formamide, and water are used alone or combined as plasticizers to enhance the dispersion and the interfacial affinity in TPS/PLA blends. In the presence of water and other plasticizers including glycerol, sorbitol, urea, and formamide, the strong intermolecular and intramolecular hydrogen bonds in starch can be weakened.

To improve the compatibility between PLA and starch, suitable compatibilizer should be added.

Besides, gelatinization of starch is also a good method to enhance the interfacial affinity. Starch is gelatinized to disintegrate granules and overcome the strong interaction of starch molecules in the presence of water and other plasticizers, which leads to well dispersion. The glass transition temperature and mechanical properties of TPS/PLA blend depend on its composition and the content of plasticizer as well (Table 3.1), indicating the compatibility between PLA and TPS is low but some degree of interaction is formed.

**Table 3.1.** Thermal and mechanical properties of thermoplastic starch/polylactide (TPS/PLA) blends.

| Content of TPS [wt%] | $T_g$ [°C] | | Tensile Strength [MPa] | Elongation at Break [%] |
|---|---|---|---|---|
| | PLA | TPS | | |
| 100 (TPS1)[a] | – | 10 | 3.4 | 152.0 |
| 90 (TPS1) | 47 | NF[b] | 2.9 | 48.8 |
| 75 (TPS1) | 53 | NF | 4.8 | 5.7 |
| 100 (TPS2)[a] | – | 43 | 19.5 | 2.8 |
| 90 (TPS2) | NF | NF | 14.1 | 1.3 |
| 75 (TPS2) | NF | NF | 12.0 | 0.9 |
| 0 | 58 | – | 68.4 | 9.4 |

[a]the content of glycerol and water in TPS1 and PTS2 are 18 and 12, 10 and 16 wt% respectively
[b]Tg value is not found in the literature

The PCL is another important member of synthetic biodegradable polymer family. It is linear, hydrophobic, partially crystalline polyester, and can be slowly degraded by microbes. Blends between starch and PCL have been well documented in the literatures. The weakness of pure starch materials including low resilience, high moisture sensitivity, and high shrinkage has been overcome by adding PCL to starch matrix even at low PCL concentration. Blending with PCL, the impact resistance and the dimensional stability of native starch is improved significantly. The glass transition temperature and mechanical properties of TPS/PCL blend are varied with its composition and the content of plasticizer (Table 3.2). As can be seen, TPS/PCL blend is similar to TPS/PLA blend in both the compatibility and the role of components.

Table 3.2. Thermal and mechanical properties of thermoplastic starch/polylactide (TPS/PCL) blends.

| Content of TPS [wt%] | $T_g$ [°C] | | Tensile Strength [MPa] | Elongation at Break [%] |
|---|---|---|---|---|
| | PCL | TPS | | |
| 100 (TPS1)[a] | – | 8.4 | 3.3 | 126.0 |
| 75 (TPS1) | 31.0 | | 5.9 | 62.6 |
| 100 (TPS2)[a] | – | 43.4 | 21.4 | 3.8 |
| 75 (TPS2) | 41 | | 10.5 | 2.0 |
| 60 (TPS2) | NF[b] | | 9.0 | 2.4 |
| 0 | –61.5 | – | 14.2 | >550.0 |

[a]the content of glycerol and water in TPS1 and PTS2 are 18 and 12, 10 and 16 wt% respectively
[b]$Tg$ value is not found in the literature

The PCL/starch blends can be further reinforced with fiber and nano-clay, respectively. Moreover, the other properties of the blends such as hydrolytic stability, degradation rate, and compatibilization between PCL and starch are also improved. PVA is a synthetic water-soluble and biodegradable polymer. PVA has excellent mechanical properties and compatibility with starch. PVA/starch blend is assumed to be biodegradable since both components are biodegradable in various microbial environments. The biodegradability of blends consisting of starch, PVA, glycerol, and urea is performed by bacteria and fungi isolated from the activated sludge of a municipal sewage plant and landfill, which indicate that microorganisms consumed starch and the amorphous region of PVA as well as the plasticizers. Meanwhile, the blend is expected to exhibit good mechanical and process properties. Owing to the strong interaction among hydroxyl groups on PVA and starch chains, all the $Tg$ of the starch/PVA blends of different compositions are lower than that of PVA. The excellent compatibility of two components make the tensile strength of the blend increases with increasing PVA concentration, and the elongation at break of the blend is almost kept constant [110–123].

In addition, PVA can be used to enhance the compatibility of starch/PLA blends. Because both starch and PVA are polyols, starch will form continuous phase with PVA during blending. As a result, the mechanical properties of the starch/PLA blends are improved in the presence of PVA. As for the blend system without PVA, starch acts as filler in the PLA continuous matrix. PLA acts as the main load-bearing phase because of the weak interaction between starch and PLA.

### Blend with biopolymers
Natural polymers such as chitosan and cellulose and their derivatives are inherently biodegradable, and exhibit unique properties. A number of investigations have been devoted to study the blend of them with starch. Starch and chitosan are abundant

naturally occurring polysaccharide. Both of them are cheap, renewable, non-toxic, and biodegradable. The starch/chitosan blend exhibits good film forming property, which is attributed to the inter- and intramolecular hydrogen bonding that formed between amino groups and hydroxyl groups on the backbone of two components. The mechanical properties, water barrier properties, and miscibility of biodegradable blend films are affected by the ratio of starch and chitosan. Extrusion of the mixture of corn starch and microcrystalline cellulose in the presence or absence of plasticizers (polyols) is used to produce edible films. By increasing the content of the cellulose component, the rupture strength is increased, whereas the elongation at break and the permeability of films for water vapor are decreased. Starch can form thermodynamically compatible blend films with water-soluble CMC when the starch content is below 25 mass%. Such films are biodegradable in presence of microorganisms. Starch-based nanocomposite film is obtained by casting the mixture of PS and flax cellulose nanocrystals. The mechanical properties and water resistance are greatly improved. The tensile strength of nanocomposite and unreinforced films are 498.2 and 11.9 MPa, respectively.

*Chemical derivatives*

One problem for starch-based blends is that starch and many polymers are non-miscible, which leads to the mechanical properties of the starch/polymer blends generally become poor. Thus, chemical strategies are taken into consideration. Chemical modifications of starch are generally carried out via the reaction with hydroxyl groups in the starch molecule. The derivatives have physicochemical properties that differ significantly from the parent starch but the biodegradability is still maintained. Consequently, substituting the hydroxyl groups with some groups or chains is an effective means to prepare starch-based materials for various needs. Graft copolymerization is an often used powerful means to modify the properties of starch. Moreover, starch-g-polymer can be used as an effective compatibilizer for starch-based blends. PCL and PLA are chemically bonded onto starch and can be used directly as thermoplastics or compatibilizer. The graft-copolymers starch-g-PCL and starch-g-PLA can be completely biodegraded under natural conditions and exhibit improved mechanical performances. To introduce PCL or PLA segments onto starch, the ring opening graft polymerization of ε-caprolactone or L-lactide with starch is carried out. Starch-g-poly(vinyl alcohol) can be prepared via the radical graft copolymerization of starch with vinyl acetate and then the saponification of the starch-g-poly(vinyl acetate). Starch-g-PVA behaves good properties of both components such as processability, hydrophilicity, biodegradability, and gelation ability.

Starch can be easily transformed into an anionic polysaccharide via chemical functionalization. For instance, a carboxylic derivative of starch, maleic starch half-ester acid (MSA), has been prepared via the esterification of starch with maleic anhydride in the presence of pyridine. MSA is an anionic polyelectrolyte, consequently it can perform ionic self-assembly with chitosan in aqueous solution and forms a polysaccharide-based polyelectrolyte complex.

**Thermoplastic Starch–clay Nanocomposites and their Characteristics**

Biodegradable polymers such as starch, poly(lactide) and poly(3-caprolactone), have attracted considerable attention in the packaging industry. Starch is a promising raw material because of its annual availability from many plants, its rather excessive production with regard to current needs and its low cost. It is known to be completely degradable in soil and water and can promote the biodegradability of a non-biodegradable plastic when blended. Starch is commonly pretreated with a plasticizer to make it thermoplastic thus enabling melt-processing. However, TPS alone often cannot meet all the requirements of a packaging material and an environmentally acceptable filler is called for to improve theproperties of TPS in such applications. Clay is a potential filler; itself a naturally abundant mineral that is toxin-free and can be used as one of the components for food, medical, cosmetic, and healthcare recipients. TPS reinforced by clay has recently been investigated. To the authors' knowledge, there are just four publications describing this new class of materials, that is TPS–clay nanocomposites. Starch is hydrophilic and forms nanocomposites with natural smectite clays and conventional composites with kaolinite. It has been shown that the tensile strength of TPS was increased from 2.6 to 3.3 MPa with the presence of 5 wt% sodium montmorillonite, while the elongation at break was increased from 47 to 57%. Also the relative water vapor diffusion coefficient of TPS was decreased to 65% and the temperature at which the composite lost 50% mass was increased from 305 to 336°C.

### Synthesis and Characterization of Chitosan-carboxymethyl Starch Hydrogels as Nano Carriers for Colon-specific Drug Delivery

Although, colon delivery has become a widely accepted route of administration of therapeutic drugs, the gastrointestinal tract presents several formidable barriers to drug delivery. Colonic drug delivery has gained increased importance not just for the delivery of the drugs for the treatment of local diseases associated with the colon but also for its potential for the delivery of proteins and therapeutic peptides. To achieve successful colonic delivery, a drug needs to be protected from absorption and/or the environment of the upper gastrointestinal tract (GIT) and then be abruptly released into the proximal colon, which is considered the optimum site for colon targeted delivery of drugs. Colon targeting is naturally of value for the topical treatment of diseases of colon such as Chron's diseases, ulcerative colitis, colorectal cancer, and amebiasis. Peptides, proteins, oligonucleotides, and vaccines pose potential candidature for colon targeted drug delivery. The novel intestinal specific drug delivery system with pH-sensitive swelling and drug release properties was developed. The pH-sensitive hydrogel containing ibuprofen pendents was used as colon-specific drug delivery. The carboxyl group of ibuprofen was converted to a vinyl ester group by reacting ibuprofen and vinyl acetate as an acylating agent in the presence of catalyst. The glucose-6-acrylate-1, 2, 3, 4-tetraacetate (GATA) monomer was prepared under mild conditions. Cubane-1, 4-dicarboxylic acid (CDA) linked to two 2-hydroxyethyl methacrylate (HEMA) group was used as the cross-linking agent (CA). The goal of oral insulin delivery devices is to protect the sensitive drug from proteolytic degradation in the stomach and upper portion of the small intestine. pH-responsive, poly(methacrylic g-ethylene glycol) hydrogels are as oral delivery vehicles for insulin. Insulin was loaded into polymeric microspheres and administered orally to healthy and diabetic Wistar rats.

In the acidic environment of the stomach, the gels were unswollen due to the formation of intermolecular polymer complexes. The insulin remained in the gel and was protected from proteolytic degradation. In the basic and neutral environments of the intestine, the complexes dissociated which resulted in rapid gel swelling and insulin release. Cross-linked 2-hydroxyethyl methacrylate (HEMA) and methacrylic acid (MAA) copolymer hydrogels are new oral delivery system for insulin. Tere-phthalic acid covalently links with 2-hydroxyethyl methacrylate (HEMA), abbreviated as cross-linking agent (CA). Free radical copolymerization of HEMA and MAA with terephthetalicacid (CA) (2, 4, and 6%) as cross-linking agent carries out at 70°C. Polymer bonded drug usually contain one solid drug bonded together in a matrix of a solid polymeric binder. They can be produced by polymerization of a monomer such as methacrylic acid, mixed with a particulate drug, by means of a chemical polymerization catalyst, such as bis-acrylamide as a cross-linking agent and ammonium persulfate as an initiator or by means of high-energy radiation, such as X-ray or gamma rays. Copolymers of 2-hydroxyethyl methacrylate and methacrylic acid are hydrogels as an important hydrogel drugdelivery system.

Radiation copolymerization of 2-hydroxyethyl methacrylate and methacrylic acid (mixed with 3, 3'- azobis(6-hydroxy benzoic acid)(ABHB) as an azo derivative of 5-aminosalicylic acid carries out with various amounts of methacryloyloxyethyl esters of terephthalic acid for cross-linking. Starch is the most abundant, renewable bio-polymer, which is very promising raw material, available at low cost for preparing of various functional polymers. Carboxymethyl starch (CMS) widely used in pharmaceuticals; however, it may need to be further modified for some special applications. Among diverse approaches that are possible for modifying polysaccharides, grafting of synthetic polymer is a convenient method for adding new properties to a polysaccharide with minimum loss of its initial properties. Graft copolymerization of vinyl monomers onto polysaccharides using free radical initiation, has attracted the interest of many scientists. Up to now, considerable works have been devoted to the grafting of vinyl monomers onto the substrates, especially starch and cellulose. Existence of polar functionally groups as carboxylic acid need not only for bioadhesive properties but also for pH-sensitive properties of polymer. Because the increase of methacrylic acid content in the hydrogels provides more hydrogen bonds at low pH and more electrostatic repulsion at high pH. Also, chitosan is a functional linear polymer derived from chitin, the most abundant natural polysaccharide on the earth after cellulose, and it is not digested in the upper GI tract by human digestive enzymes. Chitosan is a copolymer consisting of 2-amino-2-deoxy-D-glucose and 2-acetamido-2-deoxy-D-glucose units links with b-(1-4) bonds. It should be susceptible to glycosidic hydrolysis by microbial enzymes in the colon because it possesses glycosidic linkages similar to those of other enzymatically depolymerized polysaccharides. Among diverse approaches that are possible for modifying polysaccharides, grafting of synthetic polymer is a convenient method for adding new properties to a polysaccharide with minimum loss of its initial properties.

It is as a part of international research program on CS/CMS modification to prepare materials with pH-sensitive properties for uses as colon-specific drug delivery. The free radical graft copolymerization of polymethacrylic acid onto CS/CMS was carried out in DMF/H2O solvent system at 40°C. The 5-aminosalicylic acid (5-ASA) is an active ingredient of agents used for the long-term maintenance therapy to prevent relapses of Crohn's disease and ulcerative colitis. However, when 5-ASA is administered orally, a large amount of the drug is absorbed from the upper gastrointestinal tract (GIT), and causes systemic side effects. Therefore, it is preferable to deliver the drug site-specifically to the colon. Research on the colonic drug delivery system can be performed in the stomach, small intestine, cecum, and colon areas. The CS/CMS/PMA–5-ASA conjugate have been synthesized as prodrug only for colon-specific area drug delivery system. The mixture modified hydrogel and 5-aminosalicylic acid (5-ASA) and salicylic acid (SA) as model drugs were converted to nano by freeze-drying method. The equilibrium swelling studies and *in vitro* release profiles were carried out in enzyme-free simulated gastric and intestinal fluids (SGF and SIF, respectively).The influences of different factors, such as content of methacrylic acid in the feed monomer and swelling have been studied.

## Modification of Carboxymethyl Starch as Nano Carriers for Oral Drug Delivery

Nano carriers have important potential applications for the administration of therapeutic molecules. The research in this area is being carried out all over the world at a great pace. Research areas cover novel properties that have been developed increased efficiency of drug delivery, improved release profiles, and drug targeting. Although oral delivery has become a widely accepted route of administration of therapeutic drugs, the gastrointestinal tract presents several formidable barriers to drug delivery. To achieve successful colonic delivery, a drug needs to be protected from absorption of the environment of the upper gastrointestinal tract (GIT) and then be abruptly released into the proximal colon, which is considered the optimum site for colon-targeted delivery of drugs. One strategy for targeting orally administered drugs to the colon includes coating drugs with pH-sensitive hydrogels. Polymer bonded drug usually contain one solid drug bonded together in a matrix of a solid polymeric binder. They can be produced by polymerizing a monomer such as methacrylic acid (MAA), mixed with a particulate drug, by means of a chemical polymerization catalyst, such as AIBN or by means of high-energy radiation, such as X-ray or gamma rays.

Natural polymers have potential pharmaceutical applications because of their low toxicity, biocompatibility, and excellent biodegradability. Starch is the most abundant, renewable biopolymer, which is very promising raw material, available at low cost for preparing of various functional polymers. The CMS widely used in pharmaceuticals; however, it may need to be further modified for some special applications. Among diverse approaches that are possible for modifying polysaccharides, grafting of synthetic polymer is a convenient method for adding new properties to a polysaccharide with minimum loss of its initial properties. Graft copolymerization of vinyl monomers onto polysaccharides using free radical initiation, has attracted the interest of many scientists. Up to now, considerable works have been devoted to the grafting of vinyl monomers onto the substrates, especially Starch and cellulose. Existence of polar functionally groups as carboxylic acid need not only for bioadhesive properties but also for pH-sensitive properties of polymer. Because the increase of MAA content in the hydrogels provides more hydrogen bonds at low pH and more electrostatic repulsion at high pH.

A research program on CMS modification to prepare materials with pH-sensitive properties for uses as colon-specific drug delivery. The free radical graft copolymerization poly methacrylic acid onto CMS was carried out at 70°C, bis-acrylamide as a cross-linking agent and persulfate as an initiator. The mixture modified hydrogel and 5-aminosalicylic acid (5-ASA) and salicylic acid (SA) as model drugs were converted to nano by freeze-drying method. The equilibrium swelling studies and *in vitro* release profiles were carried out in enzyme-free simulated gastric and intestinal fluids (SGF and SIF, respectively). The influences of different factors, such as content of MAA in the feed monomer and swelling were studied.

## SYNTHESIS AND CHARACTERIZATION OF NEW ELECTRORHEOLOGICAL FLUIDS BY CARBOXYMETHYL STARCH NANOCOMPOSITES

Electrorheological fluids are typical materials from a microsized particles with di-electrical good properties. The characterization of electrorheological fluids such as viscosity, yield stress, and shear modulus can change in the different. This characterization find practical applications in many fields, for example, shock absorbers, active devices, human muscle simulators, photonic crystal, and various control systems.

Feldspar comprises a group of minerals containing potassium, sodium, calcium, and aluminum silicates. They are the most common rock-forming minerals. The common feldspar is potassium feldspar, namely, orthoclase ($K_2O$, $Al2O_3$, $6SiO_2$). Sodium feldspar is albite ($Na_2O$, $Al_2O_3$, $6SiO_2$) and calcium feldspar is anorthite (CaO, $Al2O_3$, $2SiO_2$). A variety of crossed, hatched, twinned orthoclase (to be seen under the petrological microscope only) is called microcline. Sodium and calcium feldspars form an isomorphous mixture known as plagioclase feldspars. In between sodium and calcium, the other feldspars of the plagioclase series are oligoclase, andesine, labradorite, and bytownite. They are composed of suitable proportions of sodium and calcium with an increasing percentage of calcium beginning from mineral oligoclase to bytownite, turning completely into calcium feldspar (anorthite). A rock containing only plagioclase feldspars is called anorthosite. The commercial feldspar is orthoclase. The potassium molecule is replaced by sodium to some extent and hence, orthoclase feldspar usually contains a small percentage of sodium. The composition range of the commercial feldspar varies within the limits of potash, soda, and up to oligoclase. Potash and soda feldspar occur as essential constituents of granite, syenite, and gneisses. However, workable deposits are found in pegmatite veins consisting mainly of feldspar, quartz-feldspar veins and also occur with mica pegmatites.

Feldspar is of widespread occurrence and is mined in almost all countries. The choice of intercalation feldspar with dimethylsulfoxide (dielectric constant is about 47) is aimed at modifying the dielectric and polarization properties of feldspar, so as to improve its electrorheological activity, reduce cost and attain the high cost performance. The feldspar/dimethyl sulfoxide/CMS nanocomposite is fabricated according to the physical and chemical design of the electrorheological fluid material. The polar liquid (DMSO) is directly intercalated into the interlayer of feldspar and then the intercalated complex is interacted with CMS by the solution method. The dielectric and conductivity properties of these nanocomposites are improved enormously. The experimental results show that by the design and control of the molecular chemical structure, the physical design for dielectric properties is achieved and thus the characterization of nanocomposite is optimized.

**KEYWORDS**

- **Coreshell structure**
- **Electrorheological fluid**
- **Polymerization**
- **Polysaccharides**

# Chapter 4

## Updates on Lamination of Nanofiber

M. Kanafchian and A.K. Haghi

---

### INTRODUCTION

Clothing is a person's second skin, since it covered great parts of the body and having a large surface area in contact with the environment. Therefore, clothing is proper interface between environment and human body, and could act as an ideal tool to enhance personal protection. Over the years, growing concern regarding health and safety of persons in various sectors, such as industries, hospitals, research institutions, battlefields, and other hazardous conditions has led to intensive research and development in field of personal protective clothing. Nowadays, there are different types of protective clothing. The simplest and most preliminary of this equipment is made from rubber or plastic that is completely impervious to hazardous substances, air, and water vapor. Another approach to protective clothing is laminating activated carbon into multilayer fabric in order to absorb toxic vapors from environment and prevent penetration to the skin [1]. The use of activated carbon is considered only a short term solution because it loses its effectiveness upon exposure to sweat and moisture. The use of semi-permeable membranes as a constituent of the protective material is another approach. In this way, reactive chemical decontaminants encapsulates in microparticles [2] or fills in microporous hollow fibers [3]. The microparticle or fiber walls are permeable to toxic vapors, but impermeable to decontaminants, so that the toxic agents diffuse selectively into them and neutralize. All of these equipments could trap such toxic pollutions but usually are impervious to air and water vapor, and thus retain body heat. In other words, a negative relationship always exists between thermal comfort and protection performance for currently available protective clothing. For example, nonwoven fabrics with high air permeability exhibit low barrier performance, whereas microporous materials, laminated fabrics, and tightly constructed wovens offer higher level of protection but lower air permeability. Thus, there still exists a very real demand for improved protective clothing that can offer acceptable levels of impermeability to highly toxic pollutions of low molecular weight, while minimizing wearer discomfort and heat stress [4].

Electrospinning provides an ultrathin membrane-like web of extremely fine fibers with very small pore size and high porosity, which makes them excellent candidates for use in filtration, membrane, and possibly protective clothing applications. Preliminary investigations have indicated that the using of nanofiber web in protective clothing structure could present minimal impedance to air permeability and extremely efficiency in trapping aerosol toxic pollutions. Potential of electrospun webs for future protective clothing systems has been investigated [5–7]. Schreuder-Gibson et al. has shown an enhancement of aerosol protection via a thin layer of electrospun fibers.

They found that the electrospun webs of nylon 66, polybenzimidazole, polyacrylo-nitrile (PAN), and polyurethane provided good aerosol particle protection, without a considerable change in moisture vapor transport or breathability of the system [5]. While nanofiber webs suggest exciting characteristics, it has been reported that they have limited mechanical properties [8, 9]. To compensate this drawback in order to use of them in protective clothing applications, electrospun nanofiber webs could be laminated via an adhesive into a multilayer fabric system [10, 11]. The protective clothing made of this multilayer fabric will provide both protection against toxic aerosol and thermal comfort for user.

The adhesives in the fabric lamination are as solvent/water-based adhesive or as hot-melt adhesive. At the first group, the adhesives are as solution in solvent or water, and solidify by evaporating of the carrying liquid. Solvent-based adhesives could "wet" the surfaces to be joined better than water-based adhesives, and also could solidify faster. But unfortunately, they are environmentally unfriendly, usually flammable and more expensive than those. Of course, it does not mean that the water-based adhesives are always preferred for laminating, since in practice, drying off water in terms of energy and time is expensive too. Besides, water-based adhesives are not resisting to water or moisture because of their hydrophilic nature. At the second group, hot-melt adhesives are environmentally friendly, inexpensive, require less heat and energy, and so are now more preferred. Generally there are two procedures to melt these adhesives; static hot-melt laminating that accomplish by flat iron or Hoffman press and continuous hot-melt laminating that uses the hot calendars. In addition, these adhesives are available in several forms; as a web, as a continuous film, or in powder form. The adhesives in film or web form are more expensive than the corresponding adhesive powders. The web form are discontinuous and produce laminates which are flexible, porous, and breathable, whereas, continuous film adhesives cause stiffening and produce laminates which are not porous and permeable to both air and water vapor. This behavior attributed to impervious nature of adhesive film and its shrinkage under the action of heat [12]. Thus, the knowledge of laminating skills and adhesive types is very essential to producing an appropriate multilayer fabric. Specifically, this subject becomes more highlight as we will laminate the ultrathin nanofiber web into multilayer fabric, because the laminating process may be adversely influenced on the nanofiber web properties. Lee et al. [7], without disclosure of laminating details, reported that the hot-melt method is more suitable for nanofiber web laminating. In this method, laminating temperature is one of the most effective parameters. Incorrect selection of this parameter may lead to change or damage ultrathin nanofiber web. Therefore, it is necessary to find out a laminating temperature which has the least effect on nanofiber web during process.

The purpose of this study is to consider the influence of laminating temperature on the nanofiber web/multilayer fabric properties to make protective fabric which is resistance against aerosol pollutions. Multilayer fabrics were made by laminating of nanofiber web into cotton fabric via hot-melt method at different temperatures. Effects of laminating temperature on the nanofiber web morphology, air transport properties, and the adhesive force were discussed.

## EXPERIMENTAL

### Electrospining and Laminating Process

The electrospinning conditions and layers properties for laminating are summarized in Table 4.1. PAN of 70,000 g/mol molecular weight from Polyacryl Co. (Isfehan, Iran) has been used with N,N-dimethylformamide (DMF) from Merck, to form a 12% Wt polymer solution after stirring for 5 h and exposing for 24 h at ambient temperature. The yellow and ripened solution was inserted into a plastic syringe with a stainless steel nozzle and then it was placed in a metering pump from World Precision Instruments (Florida, USA). Next, this set installed on a plate which it could traverse to left–right direction along drum collector (Fig. 4.1). The electrospinning process was carried out for 8 h and the nanofibers were collected on an aluminum-covered rotating drum which was previously covered with a Poly-propylene Spun-bond Nonwoven (PPSN) substrate. After removing of PPSN covered with nanofiber from drum and attaching another layer of PPSN on it, this set was incorporated between two cotton weft-warp fabrics as a structure of fabric-PPSN-nanofiber web- PPSN-fabric (Fig. 4.2). Finally, hot-melt laminating performed using a simple flat iron for 1 min, under a pressure of 9gf/cm² and at temperatures 85, 110, 120, 140, 150°C (above softening point of PPSN) to form the multilayer fabrics.

**Table 4.1.** Electrospinning conditions and layers properties for laminating.

| Electrospinning conditions | | Layer properties | |
|---|---|---|---|
| | | PPSN | |
| Polymer concentration | 12% | Thickness | 0.19mm |
| Flow rate | 1 μl/h | Air permeability | 824 cm³/s/cm² |
| | | Melting point | 140°C |
| Nozzleinner diameter | 0.4 mm | Mass | 25 g/m² |
| Nozzle-Drum distance | 7 cm | Nanofiber web Mass | 3.82 g/m² |
| Voltage | 11 KV | Fabric | |
| Drum speed | 9 m/min | Thickness | 0.24 mm |
| Spinning Time | 8 hr | Warp-weft density | 25×25 per cm |

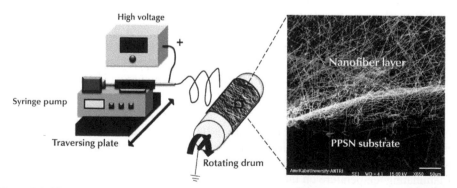

**Figure 4.1.** Electrospinning setup and an enlarged image of nanofiber layer on PPSN.

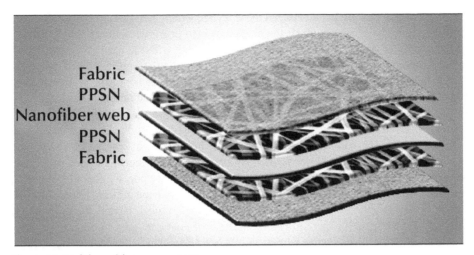

**Figure 4.2.** Multilayer fabric components.

## Nanofiber Web Morphology

A piece of multilayer fabrics were freeze fractured in liquid nitrogen and after sputter-coating with Au/Pd, a cross-section image of them captured using a scanning electron microscope (Seron Technology, AIS-2100, Korea).

Also, to consider the nanofiber web surface after hot-melt laminating, other laminations were prepared by a non-stick sheet made of Teflon (0.25 mm thickness) as a replacement for one of the fabrics (fabric/PPSN/nanofiber web/PPSN/Teflon sheet). Laminating process was carried out at the same conditions which mentioned to produce primary laminations. Finally, after removing of Teflon sheet, the nanofiber layer side was observed under an optical microscope (Microphot-FXA, Nikon, Japan) connected to a digital camera.

## Measurement of Air Permeability

Air permeability of multilayer fabrics after laminating were tested by fabric air permeability tester (Textest FX3300, Zürich, Switzerland). It was tested five pieces of each sample under air pressure of 125 Pa, at ambient condition (16°C, 70% RH) and then obtained average air permeability.

## RESULTS AND DISCUSSION

In electrospinning phase, PPSN was chosen as a substrate to provide strength to the nanofiber web and to prevent of its destruction in removing from the collector. In Fig. 4.1, an ultrathin layer of nanofiber web on PPSN layer is illustrated, which conveniently shows the relative fiber sizes of nanofibers web (approximately 380 nm) compared to PPSN fibers. Also, the figure shows that the macropores of PPSN substrate is covered with numerous electrospun nanofibers, which will create innumerable microscopic pores in this system. But in laminating phase, this substrate acts as an adhesive and causes to bond the nanofiber web to the fabric. In general, it is relatively simple to create a strong bond between these layers, which guarantees no delamination or failure in multilayer structures; the challenge is to preserve the original properties of the nanofiber web and fabrics to produce a laminate with the required appearance, handle, thermal comfort, and protection. In other words, the application of adhesive should have minimum affect on the fabric flexibility or on the nanofiber web structure. In order to achieve to this aim, it is necessary that: (a) the least amount of a highly effective adhesive applied, (b) the adhesive correctly cover the widest possible surface area of layers for better linkage between them, and (c) the adhesive penetrate to a certain extent of the nanofiber web/fabric [12]. Therefore, we selected PPSN, which is a hot-melt adhesive in web form. As mentioned above, the perfect use of web form adhesive can be lead to produce multilayer fabrics which are porous, flexible, and permeable to both air and water vapor. On the other hand, since the melting point of PPSN is low, hot-melt laminating can perform at lower temperatures. Hence, the probability of shrinkage that may happen on layers in effect of heat becomes smaller. Of course in this study, we utilized cotton fabrics and PAN nanofiber web for laminating, which intrinsically are resistant to shrinkage even at higher temperatures (above laminating temperature). By this description, laminating process performed at five different temperatures to consider the effect of laminating temperature on the nanofiber web/multilayer fabric properties.

Figure 4.3(A–E) shows a SEM image of multilayer fabric cross-section after laminating at different temperatures. It is obvious that these images do not deliver any information about nanofiber web morphology in multilayer structure, so it becomes impossible to consider the effect of laminating temperature on nanofiber web. Therefore, in a novel way, we decided to prepare a secondary multilayer by substitution of one of the fabrics (ref. Fig. 4.2) with Teflon sheet. By this replacement, the surface of nanofiber web will become accessible after laminating; because Teflon is a non-stick material and easily separates from adhesive.

Figure 4.3(a–e) presents optical microscope images of nanofiber web and adhesive after laminating at different temperatures. It is apparent that the adhesive gradually flattened on nanofiber web (Fig. 4.3(a–c)) when laminating temperature increased to melting point of adhesive (140°C). This behavior is attributed to increment in plasticity of adhesive because of temperature rise and the pressure applied from the iron weight. But, by selection of melting point as laminating temperature, the adhesive completely melted and began to penetrate into the nanofiber web structure instead of spread on it (Fig. 4.3(d)). This penetration, in some regions, was continued to some extent that the adhesive was even passed across the web layer. The dark crisscross lines in Fig. 4.3(d) obviously show where this excessive penetration is occurred. The adhesive penetration could intensify by increasing of laminating temperature above melting point; because the fluidity of melted adhesive increases by temperature rise. Figure 4.3(e) clearly shows the amount of adhesive diffusion in the web which was laminated at 150°C. At this case, the whole diffusion of adhesive lead to create a transparent film and to appear the fabric structure under optical microscope.

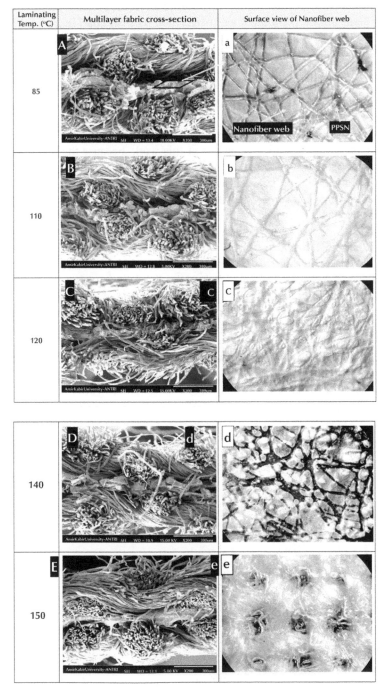

**Figure 4.3.** SEM images of multilayer fabric cross-section at 200 magnification (A–E) and optical microscope images of nanofiber web surface at 100 magnification (a–e).

Also, to examine how laminating temperature affect the breathability of multilayer fabric, air permeability experiment was performed. The bar chart in Fig. 4.4 indicates the effect of laminating temperature on air transport properties of multilayer fabrics. As might be expected, the air permeability decreased with increasing laminating temperature. This procedure means that the air permeability of multilayer fabric is related to adhesive's form after laminating, because the PAN nanofiber web and cotton fabrics intrinsically are resistant to heat (ref. Fig. 4.3). Of course, it is to be noted that the pressure applied during laminating can leads to compact the web/fabric structure and to reduce the air permeability too. Nevertheless, this parameter did not have effective role on air permeability variations at this work, because the pressure applied for all samples had the same quantity. As discussed, by increasing of laminating temperature to melting point, PPSN was gradually flattened between layers so that it was transformed from web-form to film-like. It is obvious in Fig. 4.3(a–c) that the pore size of adhesive layer becomes smaller in effect of this transformation. Therefore, we can conclude that the adhesive layer as a barrier resists to convective air flow during experiment and finally reduces the air permeability of multilayer fabric according to the pore size decrease. But, this reason was not acceptable for the samples that were laminated at melting point (140°C); since the adhesive was missed self layer form because of penetration into the web/fabric structures (Fig. 4.3(d)). At these samples, the adhesive penetration leads to block the pores of web/fabrics and to prevent of the air pass during experiment. It should be noted that the adhesive was penetrated into the web much more than the fabric, because PPSN structurally had more surface junction with the web (Fig. 4.3(A–E)). Therefore, at here, the nanofiber web contained the adhesive itself could form an impervious barrier to air flow.

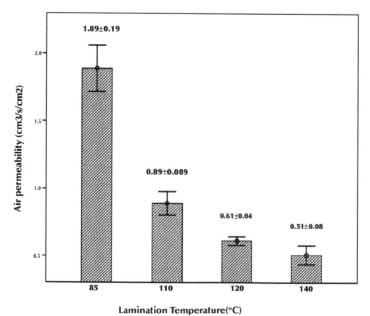

**Figure 4.4.** Air permeability of multilayer fabric after laminating at different temperatures.

Furthermore, we only observed that the adhesive force between layers was improved according to temperature rise. For example, the samples laminated at 85°C were exhibited very poor adhesion between the nanofiber web and the fabrics as much as they could be delaminated by light abrasion of thumb. Generally, it is essential that no delamination occurs during use of this multilayer fabric, because the nanofiber web might be destroyed due to abrasion of fabric layer. Before melting point, improving the adhesive force according to temperature rise is simultaneously attributed to the more penetration of adhesive into layers and the expansion of bonding area between them, as already discussed. Also at melting point, the deep penetration of adhesive into the web/fabric leads to increase in this force.

## CONCLUSION

In this study, the effect of laminating temperature on the nanofiber web/multilayer fabric properties investigated to make next-generation protective clothing. First, we demonstrate that it is impossible to consider the effect of laminating temperature on the nanofiber web morphology by a SEM image of multilayer fabric cross-section. Thus, we prepared a surface image of nanofiber web after laminating at different temperature using an optical microscope. It was observed that nanofiber web was approximately unchanged when laminating temperature was below PPSN melting point. In addition, to compare air transport properties of multilayer fabrics, air permeability tests were performed. It was found that by increasing laminating temperature, air permeability was decreased. Furthermore, it only was observed that the adhesive force between layers in multilayer fabrics was increased with temperature rise. These results indicate that laminating temperature is an effective parameter for laminating of nanofiber web into fabric structure. Thus, varying this parameter could lead to developing fabrics with different levels of thermal comfort and protection depending on our need and use. For example, laminating temperature should be selected close to melting point of adhesive, if we would produce a protective fabric with good adhesive force and medium air permeability.

## ACKNOWLEDGMENT

The authors would like to thank Dr. Shahram Arbab for preparing of multilayer fabric cross-section.

## KEYWORDS

- **Electrospinning**
- **Nanofiber web**
- **Poly-propylene Spun-bond Nonwoven**
- **Scanning Electron Microscopy image**

# Chapter 5

## Electrospinning of Chitosan (CHT)

Z. Moridi Mahdieh, V. Mottaghitalab, N. Piri, and A.K. Haghi

### INTRODUCTION

Over the recent decades, scientists interested to creation of polymer nanofibers due to their potential in many engineering and medical applications [1]. According to various outstanding properties such as very small fiber diameters, large surface area per mass ratio, high porosity along with small pore sizes and flexibility, electrospun nanofiber mats have found numerous applications in diverse areas. For example in biomedical field nanofibers plays a substantial role in tissue engineering [2], drug delivery [3], and wound dressing [4]. Electrospinning is a sophisticated and efficient method by which fibers produces with diameters in nanometer scale entitled as nanofibers. In electrospinning process, a strong electric field is applied on a droplet of polymer solution (or melt) held by its surface tension at the tip of a syringe needle (or a capillary tube). As a result, the pendent drop will become highly electrified and the induced charges distributes over its surface. Increasing the intensity of electric field, the surface of the liquid drop will be distorted to a conical shape known as the Taylor cone [5]. Once the electric field strength exceeds a threshold value, the repulsive electric force dominates the surface tension of the liquid and a stable jet emerges from the cone tip. The charged jet then accelerates toward the target and rapidly thins and dries because of elongation and solvent evaporation. As the jet diameter decreases, the surface charge density increases and the resulting high repulsive forces split the jet to smaller jets. This phenomenon may take place several times leading to many small jets. Ultimately, solidification is carried out and fibers deposits on the surface of the collector as a randomly oriented nonwoven mat [6–7]. Figure 5.1 shows a schematic illustration of electrospinning setup.

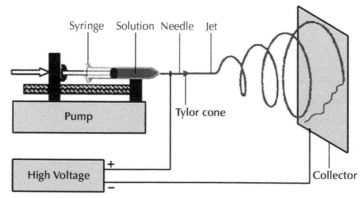

**Figure 5.1.** A typical image of electrospinning process [8].

The physical characteristics of electrospun nanofibers such as fiber diameter depend on various parameters which are mainly divided into three categories: solution properties (solution viscosity, solution concentration, polymer molecular weight, and surface tension), processing conditions (applied voltage, volume flow rate, spinning distance, and needle diameter), and ambient conditions (temperature, humidity, and atmosphere pressure) [9]. Numerous applications require nanofibers with desired properties suggesting the importance of the process control. It does not come true unless having a comprehensive outlook of the process and quantitative study of the effects of governing parameters. In this context, Sukigara et al. [10] were assessed the effect of concentration on diameter of electrospun nanofibers.

Beside physical characteristics, medical scientists showed a remarkable attention to biocompatibility and biodegradability of nanofibers made of biopolymers such as collagen [11], fibrogen [12], gelatin [13], silk [14], chitin [15], and chitosan (CHT) [16]. Chitin is the second abundant natural polymer in the world and CHT (poly-(1-4)-2-amino-2-deoxy-β-D-glucose) is the deacetylated product of chitin [17]. CHT is well known for its biocompatible and biodegradable properties [18].

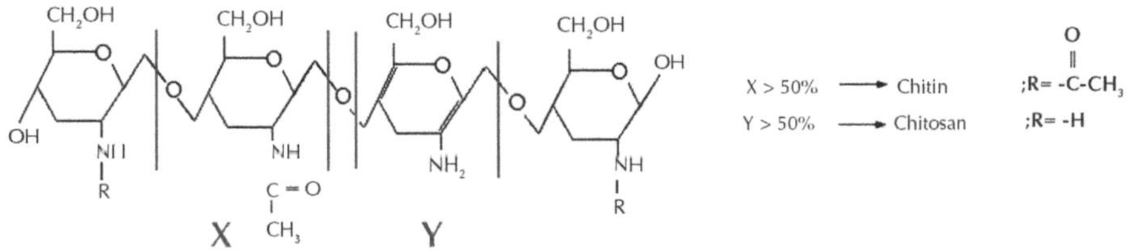

**Scheme 5.1.** Chemical structures of chitin and chitosan biopolymers.

The CHT is insoluble in water, alkali, and most mineral acidic systems. However, though its solubility in inorganic acids is quite limited, CHT is in fact soluble in organic acids, such as dilute aqueous acetic, formic, and lactic acids. CHT also has free amino groups, which make it a positively charged polyelectrolyte. This property makes CHT solutions highly viscous and complicates its electrospinning [19]. Furthermore, the formation of strong hydrogen bonds in a 3-D network prevents the movement of polymeric chains exposed to the electrical field [20].

Different strategies were used for bringing CHT in nanofiber form. The three top most abundant techniques includes blending of favorite polymers for electrospinning process with CHT matrix [21–22], alkali treatment of CHT backbone to improve electro spinnability through reducing viscosity [23] and employment of concentrated organic acid solution to produce nanofibers by decreasing of surface tension [24]. Electrospinning of Polyethylene oxide (PEO)/CHT [21] and polyvinyl alcohol (PVA)/CHT [22] blended nanofiber are two recent studies based on first strategy. In second protocol, the molecular weight of CHT decreases through alkali treatment. Solutions of the treated CHT in aqueous 70–90% acetic acid produce nanofibers with appropriate quality and processing stability [23].

Using concentrated organic acids such as acetic acid [24] and triflouroacetic acid (TFA) with and without dichloromethane (DCM) [25–26] reported exclusively for producing neat CHT nanofibers. They similarly reported the decreasing of surface tension and at the same time enhancement of charge density of CHT solution without significant effect on viscosity. This new method suggests significant influence of the concentrated acid solution on the reducing of the applied field required for electrospinning.

The mechanical and electrical properties of neat CHT electrospun natural nanofiber mat can be improved by addition of the synthetic materials including carbon nanotubes (CNTs) [27]. CNTs are one of the important synthetic polymers that were discovered by Iijima in 1991 [28]. CNTs either single walled nanotubes (SWNTs) or multiwalled nanotubes (MWNTs) combine the physical properties of diamond and graphite. They are extremely thermally conductive like diamond and appreciably electrically conductive like graphite. Moreover, the flexibility and exceptional specific surface area to mass ratio can be considered as significant properties of CNTs [29]. The scientists are becoming more interested to CNTs for existence of exclusive properties such as superb conductivity [30] and mechanical strength for various applications. To best of our knowledge, there has been no report on electrospinning of CHT/MWNTs blend, except those ones [30–31] that use PVA to improve spinnability. Results showed uniform and porous morphology of the electrospun nanofibers. Despite adequate spinnability, total removing of PVA from nanofiber structure to form conductive substrate is not feasible. Moreover, thermal or alkali solution treatment of CHT/PVA/MWNTs nanofibers extremely influence on the structural morphology and mechanical stiffness. The CHT/CNT composite can be produced by the hydrogen bonds due to hydrophilic positively charged polycation of CHT due to amino groups and hydrophobic negatively charged of CNT due to carboxyl and hydroxyl groups [32–34].

In current study, it has been attempted to produce a CHT/MWNTs nanofiber without association of processing agent to facile electrospinng process. In addition, a new approach explored to provide highly stable and homogenous composite spinning solution of CHT/MWNTs in concentrated organic acids. This in turn present, a homogenous conductive CHT scaffolds which is extremely important for biomedical implants.

## EXPERIMENTAL

### Materials
The CHT with degree of deacetylation of 85% and molecular weight of $5 \times 10^5$ was supplied by Sigma-Aldrich. The MWNTs, supplied by Nutrino, have an average diameter of 4 nm and purity of about 98%. All of the other solvents and chemicals were commercially available and used as received without further purification.

### Preparation of CHT-MWNTs Dispersions
A Branson Sonifier 250 operated at 30 W used to prepare the MWNT dispersions in CHT/organic acid (90% wt acetic acid, 70/30 TFA/DCM) solution based on different protocols. In first approach, 3 mg of as received MWNTs was dispersed into deionized water or DCM using solution sonicating for 10 min (current work, Sample 1).

Different amount of CHT was then added to MWNTs dispersion for preparation of a 8–12% wt solution and then sonicated for another 5 min. Figure 5.2 shows two different protocols used in this study.

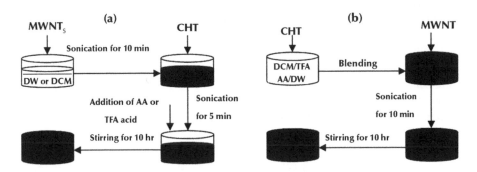

**Figure 5.2.** Two protocols used in this study for preparation of MWNTs/CHT dispersion (a) Current study and (b) Ref. [35].

In next step, organic acid solution added to obtain a CHT/MWNT solution with total volume of 5 mL and finally the dispersion was stirred for another 10 h. The Sample 2 was prepared using second technique. Same amount of MWNTs were dispersed in CHT solution, and the blend with total volume of 5 mL were sonicated for 10 min and dispersion was stirred for 10 h [35].

### Electrospinning of CHT/MWNT Dispersion

After the preparation of spinning solution, it transferred to a 5 mL syringe and became ready for spinning of nanofibers. The experiments carried out on a horizontal electrospinning setup shown schematically in Fig. 5.1. The syringe containing CHT/MWNTS solution was placed on a syringe pump (New Era NE-100) used to dispense the solution at a controlled rate. A high voltage DC power supply (Gamma High Voltage ES-30) employed to generate the required electric field for electrospinning. The positive electrode and the grounding electrode of the high voltage supplier attaches respectively to the syringe needle and flat collector wrapped with aluminum foil where electrospun nanofibers accumulates via an alligator clip to form a nonwoven mat. The voltage and the tip-to-collector distance fixed respectively on 18–24 kV and 4–10 cm. In addition, the electrospinning carried out at room temperature and the aluminum foil removed from the collector.

### Measurements and Characterizations Method

A small piece of mat placed on the sample holder and gold sputter-coated (Bal-Tec). Thereafter, the micrograph of electrospun CHT/MWNT fibers obtained using scanning electron microscope (SEM, Phillips XL-30). Fourier transform infrared spectra (FTIR) recorded using a Nicolet 560 spectrometer to investigate the interaction between CHT and MWNT in the range of 800–4000 cm$^{-1}$ under a transmission mode.

The size distribution of the dispersed solution evaluates with a Zetasizer (Malvern Instruments). The conductivity of nanofiber samples measured using a homemade four-probe electrical conductivity cell operating at constant humidity. The electrodes were circular pins with separation distance of 0.33 cm and fibers were connected to pins by silver paint (SPI). Between the two outer electrodes, a constant DC current applies by Potentiostat/Galvanostat model 363 (Princeton Applied Research). The generated potential difference between the inner electrodes and the current flow recorded by digital multimeter 34401A (Agilent). Figure 5.3 illustrates the experimental setup for conductivity measurement. The conductivity ($\delta$: S/cm) of the nanofiber thin film with rectangular surface can then be calculated according to equation 1 which parameters call for length (L:cm), width (W:cm), thickness (t:cm), DC current applied (mA), and the potential drop across the two inner electrodes (mV). All measuring repeated at least five times for each set of samples.

$$\delta = \frac{I \times L}{V \times W \times t} \qquad (1)$$

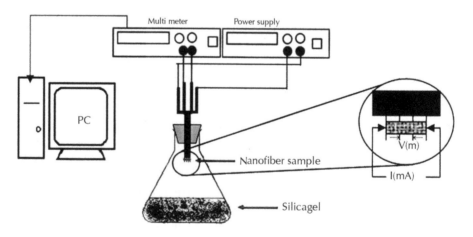

**Figure 5.3.** The experimental setup for four-probe electrical conductivity measurement of nanofiber thin film.

## RESULTS AND DISCUSSION

### The Characteristics of MWNT/CHT Dispersion

Utilization of MWNTs in biopolymer matrix initially requires their homogenous dispersion in a solvent or polymer matrix. Dynamic light scattering (DLS) is a sophisticated technique used for evaluation of particle size distribution. DLS provides many advantages for particle size analysis to measures a large population of particles in a very short time period, with no manipulation of the surrounding medium. DLS of MWNTs dispersions indicate that the hydrodynamic diameter of the nanotube bundles is between 150 and 400 nm after 10 min of sonication for Sample 2. (see Fig. 5.4)

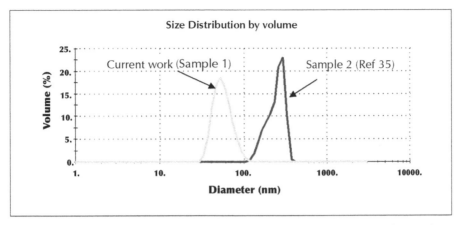

**Figure 5.4.** Hydrodynamic diameter distribution of MWNT bundles in CHT/acetic acid (1%) solution for different preparation technique.

The MWNTs bundle in Sample 1 (different approach but same sonication time compared to Sample 2) shows a range of hydrodynamic diameter between 20 and 100 nm. (Fig. 5.4).The lower range of hydrodynamic diameter for Sample 1 correlates to more exfoliated and highly stable nanotubes strands in CHT solution. The higher stability of Sample 1 compared to Sample 2 over a long period of time is confirmed by solution stability test. The results presented in Fig. 5.5 indicate that procedure employed for preparation of Sample 1 (current work) was an effective method for dispersing MWNTs in CHT/acetic acid solution. However, MWNTs bundles in Sample 2 were found to re-agglomerate upon standing after sonication, as shown in Fig. 5.5 where indicate the sedimentation of large agglomerated particles.

**Figure 5.5.** Stability of CHT-MWNT dispersions (a) Current work (Sample 1) and (b) Ref. [35] (Sample 2).

Despite the method reported in ref. 35 neither sedimentation nor aggregation of the MWNTs bundles observed in Sample 1. Presumably, this behavior in Sample 1 can be attributed to contribution of CHT biopolymer to forms an effective barrier against re-agglomeration of MWNTs nanoparticles. In fact, using sonication energy, in first step without presence of solvent, make very tiny exfoliated but unstable particle in water as dispersant. Instantaneous addition of acetic acid as solvent and long mixing most likely helps the wrapping of MWNTs strands with CHT polymer chain.

Figure 5.6 shows the FTIR spectra of neat CHT solution and CHT/MWNTs dispersions prepared using strategies explained in experimental part. The interaction between the functional group associated with MWNTs and CHT in dispersed form has been understood through recognition of functional groups. The enhanced peaks at ~1600 cm$^{-1}$ can be attributed to (N-H) band and (C = O) band of amid functional group. However, the intensity of amid group for CHT/MWNTs dispersion increases presumably due to contribution of G band in MWNTs. More interestingly, in this region, the FTIR spectra of MWNTs-CHT dispersion (Sample 1) have been highly intensified compared to Sample 2 [35]. It correlates to higher chemical interaction between acid functionalized C-C group of MWNTs and amid functional group in CHT.

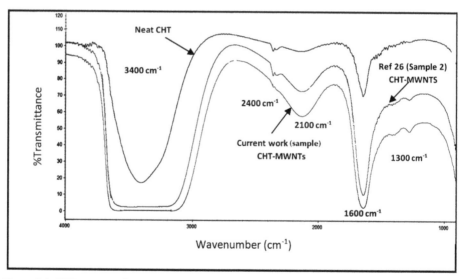

**Figure 5.6.** FTIR spectra of CHT-MWCNT in 1% acetic acid with different techniques of dispersion.

This probably is the main reason of the higher stability and lower MWNTs dimension demonstrated in Figs. 5.4 and 5.5. Moreover, the intensity of protonated secondary amine absorbance at 2400 cm$^{-1}$ for Sample 1 prepared by new technique is negligible compared to Sample 2 and neat CHT. Furthermore, the peak at 2123 cm$^{-1}$ is a characteristic band of the primary amine salt, which is associated to the interaction between positively charged hydrogen of acetic acid and amino residues of CHT. In addition, the broad peaks at ~3410 cm$^{-1}$ due to the stretching vibration of OH group superimposed on NH stretching bond and broaden due to hydrogen bonds of polysaccharides. The broadest peak of hydrogen bonds observed at 3137–3588 cm$^{-1}$ for MWNTs/CHT dispersion prepared by new technique (Sample1).

## The Physical and Morphological Characteristics of MWNTs/CHT Nanofiber

The different solvents including acetic acid 1–90%, pure formic acid, and TFA/DCM tested for preparation of spinning solution-using protocol explained for Sample 1. Upon applying of the high voltage even above 25 kV no polymer jet forms using of acetic acid 1–30% and formic acid as the solvent for CHT/CNT. However, experimental observation shows bead formation when the acetic acid (30–90%) used as the solvent. Therefore, one does not expect the formation of electrospun fiber of MWNT/CHT using prescribed solvents (data not shown).

Figure 5.7 shows SEMs of the MWNTs/CHT electrospun nanofibers in different concentration of CHT in TFA/DCM (70:30) solvent. As presented in Fig. 7(a), at low concentrations of CHT, the beads deposited on the collector and thin fibers co-exited among the beads. When the concentration of CHT increases as shown in Fig. 5.7(a–c) the bead density decreases. Figure 5.7(c) shows homogenous electrospun nanofibers with minimum beads, thin and interconnected fibers. More increasing of concentration of CHT lead to increasing of interconnected fibers at Fig. 5.7(d–e). Figure 5.8 shows the effect of concentration on average diameter of MWNTs/CHT electrospun nanofibers. Our assessments indicate that the fiber diameter of MWNTs/CHT increases with the increasing of the CHT concentration. In this context, similar results reported in previously published work [36–37]. Hence, MWNTs/CHT (10% wt) solution in TFA/DCM (70:30) considered as resulted as optimized concentration. An average diameter of 275 nm (Fig. 5.7(c): diameter distribution, 148–385) investigated for this conditions. Table 5.1 lists the variation of nanofiber diameter and four probe electrical conductivity based on the different loading of CHT. One can expect the lower conductivity, once the CHT content increases. However, the higher the CHT concentration, the thinner fiber forms. Therefore, the decreasing of conductivity at higher CHT concentration damps by decreasing of nanofiber diameter. This led to a nearly constant conductivity over entire measurements.

Table 5.1. The variation of conductivity and mean nanofiber diameter versus chitosan loading.

| % CHT (%w/v) | % MWNT (%w/v) | Voltage (kV) | Tip to collector (cm) | Diameter (nm) | Conductivity (S/cm) |
|---|---|---|---|---|---|
| 8 | 0.06 | 24 | 5 | 137 ± 58 | NA |
| 9 | 0.06 | 24 | 5 | 244 ± 61 | 9 × 10$^{-5}$ |
| 10 | 0.06 | 24 | 5 | 275 ± 70 | 9 × 10$^{-5}$ |
| 11 | 0.06 | 24 | 5 | 290 ± 87 | 8 × 10$^{-5}$ |
| 12 | 0.06 | 24 | 5 | Non-uniform | NA |

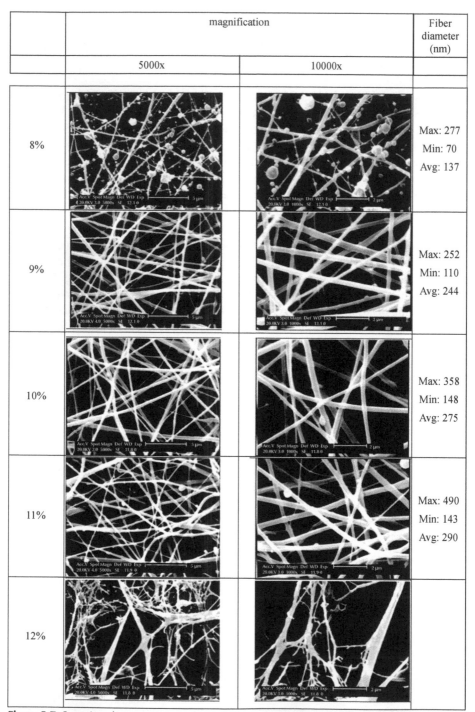

| | magnification | | Fiber diameter (nm) |
|---|---|---|---|
| | 5000x | 10000x | |
| 8% | | | Max: 277 Min: 70 Avg: 137 |
| 9% | | | Max: 252 Min: 110 Avg: 244 |
| 10% | | | Max: 358 Min: 148 Avg: 275 |
| 11% | | | Max: 490 Min: 143 Avg: 290 |
| 12% | | | |

**Figure 5.7.** Scanning electron micrographs of electrospun nanofibers at different CHT concentration (wt %): (a) 8, (b) 9, (c) 10, (d) 11, (e) 12, 24 kV, 5 cm, TFA/DCM: 70/30, (0.06% wt MWNTs).

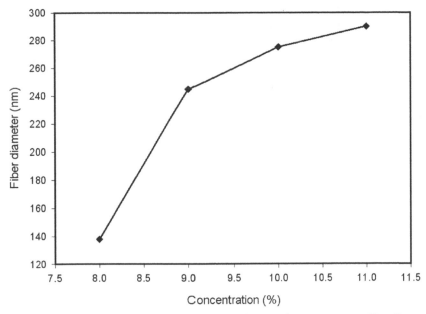

**Figure 5.8.** The effect of the CHT concentration in CHT/MWNT dispersion on nanofiber diameter.

Figure 5.9 shows the influence of voltage on morphology of CHT/MWNT electrospun nanofibers acquired SEM images. In our experiments, 18 kV attains as threshold voltage, where fiber formation occurs.

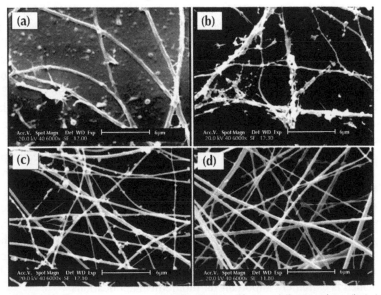

**Figure 5.9.** Scanning electronic micrographs of electrospun fibers at different voltage (kV): (a) 18, (b) 20, (c) 22, (d) 24, 5 cm, 10% wt, TFA/DCM: 70/30. (0.06%wt MWNTs).

For lower voltage, the beads and some little fiber deposited on collector (Fig. 5.9(a)). As shown in Fig. 5.9(a–d), the beads decreases while voltage increasing from 18 kV to 24 kV. The collected nanofibers by applying 18 kV (9a) and 20 kV (9b) were not quite clear and uniform. The higher the applied voltage, the uniform nanofibers with narrow distribution starts to form. The average diameter of fibers, 22 kV (9c), and 24 kV (9d), respectively, were 204 (79–391), and 275 (148–385). The conductivity measurement given in Table 5.2 confirms our observation in first set of conductivity data. As can be seen from last row, the amount of electrical conductivity reaches to a maximum level of $9 \times 10^{-5}$ at prescribed setup.

**Table 5.2.** The variation of conductivity and mean nanofiber diameter versus applied voltage.

| % CHT (%w/v) | % MWNT (%w/v) | Voltage (kV) | Tip to collector (cm) | Diameter (nm) | Conductivity (S/cm) |
|---|---|---|---|---|---|
| 10 | 0.06 | 18 | 5 | Non-uniform | NA |
| 10 | 0.06 | 20 | 5 | Non-uniform | NA |
| 10 | 0.06 | 22 | 5 | $201 \pm 66$ | $6 \times 10^{-5}$ |
| 10 | 0.06 | 24 | 5 | $275 \pm 70$ | $9 \times 10^{-5}$ |

The distance between tips to collector is another parameter that controls the fiber diameter and morphology. Figure 5.10 shows the change in morphologies of CHT/MWNTs electrospun nanofibers at different distance. When the distance is not long enough, the solvent could not find opportunity to be separated. Hence, the interconnected thick nanofiber deposits on the collector (Fig. 5.10(a)). However, the adjusting of the distance on 5 cm (Fig. 5.10(b)) lead to homogenous nanofibers with negligible beads and interconnected areas. However, the beads increases by increasing of distance of the tip-to-collector as represented from Fig. 5.10(b–f). Similar results was observed for CHT nanofibers reported by Geng et al. [24]. Also, the results show that the diameter of electrospun fibers decreases by increasing of distance tip to collector in Fig. 5.10(b), 5.10(c), 5.10(d), respectively, 275 (148–385), 170 (98–283), 132 (71–224). Similarly, the previous works reports a decreasing trend in nanofiber diameter once distance increases [38–39]. A remarkable defects and non-homogeneity appears for those fibers prepared at a distance of 8 cm (Fig. 5.10(e)) and 10 cm (Fig. 5.10(f)). However, a 5 cm distance selected as proper amount for CHT/MWNT electrospinning process. Conductivity results also are in agreement with those data obtained in previous parts (Table 5.3).

The non-homogeneity and huge bead densities plays as a barrier against electrical current and still a bead free and thin nanofiber mat shows higher conductivity compared to other samples. Experimental framework in this study was based on parameter adjusting for electrospinning of conductive CHT/MWNTs nanofiber. It can be expected that, the addition of nanotubes can boost conductivity and change morphological aspects, which is extremely important for biomedical applications.

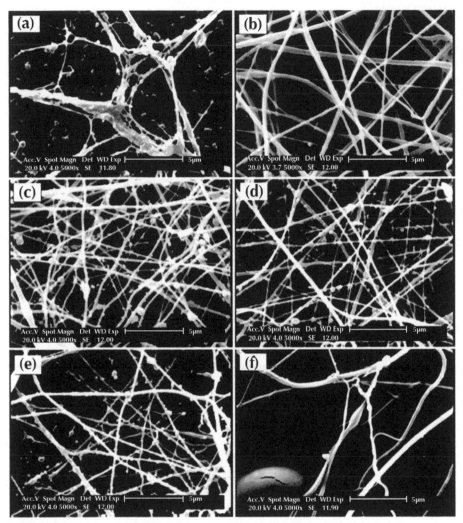

**Figure 5.10.** Scanning electronic micrographs of electrospun fibers of Chitosan/Carbon nanotubes at different tip-to-collector distances (cm): (a) 4, (b) 5, (c) 6, (d) 7, (e) 8, (f) 10, 24 kV, 10% wt, TFA/DCM: 70/30.

**Table 5.3.** The variation of conductivity and mean nanofiber diameter versus applied voltage.

| % CHT (%w/v) | % MWNT (%w/v) | Voltage (kV) | Tip to collector (cm) | Diameter (nm) | Conductivity (S/cm) |
|---|---|---|---|---|---|
| 10 | 0.06 | 24 | 4 | Non-uniform | NA |
| 10 | 0.06 | 24 | 5 | $275 \pm 70$ | $9 \times 10^{-5}$ |
| 10 | 0.06 | 24 | 6 | $170 \pm 58$ | $6 \times 10^{-5}$ |
| 10 | 0.06 | 24 | 7 | $132 \pm 53$ | $7 \times 10^{-5}$ |
| 10 | 0.06 | 24 | 8 | Non-uniform | NA |
| 10 | 0.06 | 24 | 10 | Non-uniform | NA |

## CONCLUSION

Conductive composite nanofiber of CHT/MWNTs produced using conventional electrospinning technique. A new protocol suggested for preparation of electrospinning solution, which shows much better stability and homogeneity compared to previous techniques. Several solvent including acetic acid 1–90%, formic acid, and TFA/DCM (70:30) investigated in the electrospinning of CHT/MWNTs dispersion. Using DLS and dispersion stability tests showed that the TFA/DCM (70:30) solvent is most preferred for nanofiber formation process with acceptable electrospinnability. The formation of nanofibers with conductive pathways regarding to exfoliated and interconntected nanotube strands is a breakthrough in CHT nanocomposite area. This is a significant improvement in electrospinning of CHT/CNT dispersion. It has been also observed that the homogenous nanofibers with an average diameter of 275 nm could be prepared with a conductivity of $9 \times 10^{-5}$.

## ACKNOWLEDGMENT

Authors would like to acknowledge Guilan Science and Technology Park (GSTP) for their technical and financial supports.

## KEYWORDS

- **Chitosan**
- **Dynamic light scattering**
- **Electrospun nanofibers**
- **Multiwalled nanotubes**
- **Scanning electron microscope**
- **Single walled nanotubes**

# Chapter 6

## Smart Nanofiber Based on Nylon 6,6/Polyethylene Glycol Blend

Mahdi Nouri, Javad Mokhtari, and Mohammad Seifpoor

### INTRODUCTION

Annually, total consumption of energy in the world reaches to $4.1 \times 10^{20}$ joules. By twenty-first century, the demand for energy will be triple of what it is now [1]. The increase of greenhouse gases emission and high fuel prices are the main driving forces to seek various sources of renewable energies [2]. Saving of energy in suitable forms that minimize the mismatch between supply and demand is one of the important challenges for researchers. One of the new sources of energy saving is the use of phase change materials (PCMs). PCMs are the latent heat storage materials that can produce high density of energy storage at a constant temperature, which is called phase change temperature. The phase change process can be in the forms of solid–solid, solid–liquid, solid–gas, liquid–gas, and vice versa. The solid–gas and the liquid–gas phase changes have more latent heat, but the high volume change during the phase change process is an important defect. The solid–liquid phase change has lower latent heat, but the lower change in volume during the process attracts more attention for its industrial uses [3].

Polyethylene glycol (PEG) is one of the PCM classes. It is mostly studied in solid–liquid PCMs. PEG has a melting temperature in the range of 3.2–68.7°C, as well as a very high phase change enthalpy depending on its molecular weight. In addition, it has a good biocompatibility due to its intrinsic molecular structure and therefore frequently employed in biomaterials [4]. PEGs are available in the molecular weight range of 200–35,000 [5]. The melting temperature of PEG is proportional to the molecular weight; when its molecular weight is lower than 20,000 [6]. As the molecular weight increase, the melting point increases too.

Because the PCM is melted and recrystallized repeatedly during performing the function, the potential technical defect of leakage during the phase change of PCM from one state to another could occur. In the field of textiles, encapsulation is a useful method to overcome this problem. Several authors have explained the process. For the sake of brevity, only few of them are mentioned in the list of references [7–11]. A novel method to solve the problem of PCM leakage is employing shape-stabilized or form-stable PCM, which includes a PCM and a supporting material. Nanoscience illustrates a powerful and simple way to produce this kind of PCM.

Electrospinning is a useful method for this approach. A novel class of shape-stabilized PCM has been developed via electrospinning in the form of ultrafine fibers of PCM/polymer composite [12–16]. These electrospun thermo-regulating fibers can be useful in various fields such as energy storage, biotechnology, and so forth. Briefly, in the electrospinning process, the polymer solution transfers through a spinneret using a high-voltage electrostatic field. The electrostatic force overcomes the surface tension of the solution and therefore a charged jet of the fluid is ejected from the hemispherical surface of the fluid at the tip of the capillary. The ejected jet typically develops a bending instability and it is then solidified to form fibers. The produced fibers often have a diameter in the range of nanometers. Morphology and diameter of electrospun fibers depend on technical parameters such as polymer type, concentration of the spinning solution, applied field strength, feeding rate, and so forth [17].

In this work, PEG with the molecular weight of 6000 as the PCM and nylon 6,6 as the supporting polymer are used. The effect of electrospinning parameters on the morphology and average fiber diameters and the thermal properties of the shape-stabilized PCM nanofibers are studied using scanning electron microscopy (SEM) and differential scanning calorimetry (DSC), respectively.

## EXPERIMENTAL

### Materials
Nylon 6,6 (MW 27,000) in granule was obtained from Tire Cord Co. (Iran). PEG-6000 and formic acid were purchased from Merck.

### Preparation of the Spinning Solution
To fabricate nylon 6,6 nanofibrous mat using electrospinning process and optimize the nylon 6,6 solution concentration for further use in electrospinning of nylon 6,6/PEG, nylon 6,6 in granule was dissolved in formic acid at room temperature with gentle stirring to prepare 7, 10, 12, and 15% (w/v) homogenous solutions. To fabricate nylon 6,6/PEG nanofibrous mat, PEG 6000 powder was mixed with the 15% (w/v) nylon 6,6 solution in the weight ratios of nylon 6,6:PEG 100:50, 100:70, 100:90, 100:110, 100:130, and 100:150. These solutions were used for electrospinning and evaluation of their shear viscosity.

### Electrospinning of Nylon 6,6 and Mixed Solutions
About 3 mL of each nylon 6,6 solution was put into a 5 mL syringe with a stainless steel needle (inner diameter: 0.7 mm) attached at the open end. The needle was connected to the emitting electrode of a high-voltage supply (Gamma High Voltage Research, USA), which is capable of generating DC voltage in the range of 0–30 kV. The electrospun nanofibers were collected on a target plate (aluminum foil) located at a distance of 10 cm from the syringe tip. A syringe pump (New Eva Pump System Inc., USA) was used to feed a constant amount of solution onto the tip. The output of the injection pump was 0.5 µl/min. The applied electrical potential was 15 kV at normal laboratory condition (about 22°C). Details of the used electrospinning apparatus were explained in previous work [18].

### Characterization
The FTIR spectra were recorded using Nicolet Magana IR560 Fourier-transformed infrared spectrometer (USA) in the spectra region of 2000–400 cm$^{-1}$. Shear viscosities of the spinning fluids were measured at 22°C with the aid of Brookfield Viscometer DVII+

(USA). Morphology, surface texture and dimensions of the gold sputtered electrospun nanofibers were determined using a LEO 1455VP (Germany) SEM. Fiber diameter distribution and average fiber diameter were determined using Microstructure Measurements software. Measurements of about 100 random points on the fibers were used for determining fiber diameter distribution and the average fiber diameter.

## RESULTS AND DISCUSSION

### Electrospinning of Nylon 6,6 Solutions: Optimization of the Solution Concentration

In order to optimize electrospinning solution concentration, nylon 6,6 solutions with different concentration were electrospun in the applied voltage of 15 kV. SEM micrographs of the electrospun samples are shown in Fig. 6.1. Attempt to obtain fibrous structure from the nylon 6,6 solution with a concentration of 7% (w/v) was unsuccessful; too many beads instead of nanofibers were formed. This could be attributed to the low viscosity of the solution, which avoids the formation of a stable drop at the tip of the needle. The concentration of the solution was then increased to 10% (w/v). As the concentration increased, the total number of beads decreased. Increasing concentration of the solution (12% w/v) led to lower number of beads and finally at the concentration of 15% (w/v), no beads in the nanofibrous structure appeared and the outcome was cylindrical, smooth, and continuous nanofibers. The average fiber diameter for the fibers obtained using the 7%, 10%, 12%, and 15% (w/v) solutions are 58, 79, 83, and 128 nm, respectively. Viscosity is an important rheological property to assess spinnability of the polymer solutions. Generally, there is a critical viscosity for a solution to be electrospun; below this viscosity, chain entanglements would be insufficient to stabilize the coulombic repulsion within the ejected jet, leading to the formation of sprayed droplets.

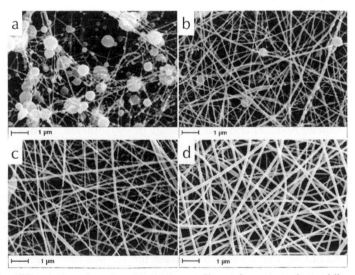

**Figure 6.1.** SEM micrographs of Nylon 6,6 nanofibers electrospun from different solution concentration: (a) 7%, (b) 10%, (c) 12%, (d) 15%.

Figure 6.2 represents shear viscosities of the nylon 6,6 solutions at various concentrations. It can be seen from Fig. 6.2, as the concentration increase, the viscosity of the solution increases, resulting in the improvement of their spinnability. At low viscosities, the solution could not provide a continuous stable jet so that the fluid stream cut, leading to so many beads on the collector instead of nanofibers as shown in Fig. 6.1. On the base of these results, the 15% concentration of the nylon 6,6 solution was selected for the further experiments. This concentration lets embedding more PEG in the defect free electrospun nanofibers.

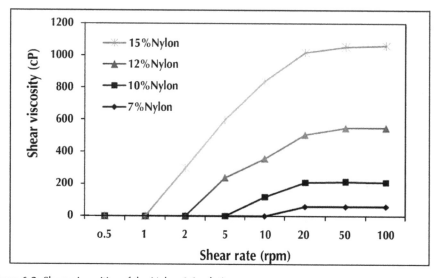

**Figure 6.2.** Shear viscosities of the Nylon 6,6 solutions.

## Electrospinning of the Blend Solutions: Effect of Blend Ratio

Electrospinning of solutions with different blend ratio was carried out in the applied voltage of 15 kV. SEM micrographs of the electrospun nanofibers are shown in Fig. 6.3. The results show that the addition of PEG to nylon 6,6 solution increases the average fiber diameter significantly. At the blend ratio of 100/50 (nylon 6,6/PEG), the nanofibers with a rounded cross section and a smooth surface were collected on the target. The diameter of nanofibers is in the range of ~40–300 nm, nanofibers with the diameter of 180 nm being the most frequently occurring. The fiber diameters have a relatively wide distribution. It may be due to splitting of the primary jet into some smaller jets. With increasing the blend ratio to 100/70 (nylon 6,6/PEG), the average fiber diameter decreased and fibers with narrower diameter distribution was obtained; a size range of ~60–260 nm with 160 nm being the most frequently occurring. Blend ratio was then increased to 100/90 (nylon 6,6/PEG), resulting in nanofibers with the diameter range of ~100–260 nm, 160 nm being the most frequently occurring. In this ratio, scattering of fiber diameters decreased but the average fiber diameter increased slightly. At the blend ratio of 100/110 (nylon 6,6/PEG), the fibers diameter was in the range of ~140–280 nm with 160 nm being the most frequently occurring. The distribution of fiber diameters decreased, but the average fiber diameter increased. With increasing the blend ratio to 100/130 (nylon 6,6/PEG), the distribution of fiber diameters increased significantly; their diameter was in the range of ~40–440 nm with 240 nm being the most frequently occurring. Although there were fibers of different diameters, they are stuck together in most samples. Partial evaporating of the solvent may be the cause of this defect. Finally, the blend ratio reached to 100/150 (nylon 6,6/ PEG). In this blend ratio, the situation was completely different. There were fibers of various diameters with so many beads on the target. Their diameter was in the range of ~60–1060 nm with 340 nm being the most frequently occurring.

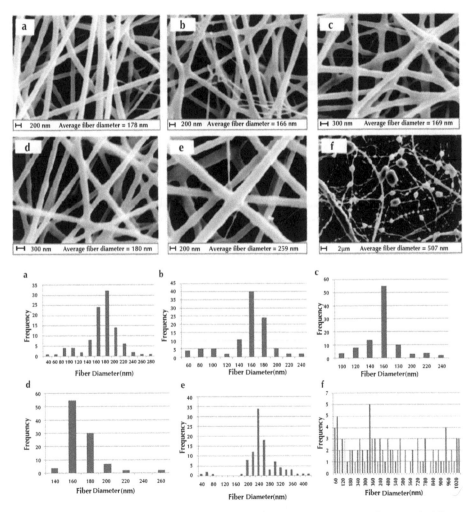

**Figure 6.3.** SEM micrographs and fiber diameter distribution of Nylon 6,6/PEG solution with different blend ratio at applied voltage of 15 kV- (a) 100/50, (b) 100/70, (c) 100/90, (d) 100/110, (e) 100/130, (f) 100/150.

Table 6.1 represents the changes in average fiber diameter and standard deviation of read fiber diameters by adding PEG to the nylon 6,6 solutions. As can be seen, the addition of PEG to the solution increases the fiber diameters. With increasing the total content of PEG in the nylon 6,6/PEG solution with different blend ratio, the average fiber diameter increases slightly till to the blend ratio of 100/110. After that, the slope of the changes increased by the blend ratio of 100/130 and at the blend ratio of 100/150 reaches to its maximum. Standard deviation of the read fiber diameters changes like average fiber diameter. Up to the blend ratio of 100/110, the changes were slight but afterward the changes were too much. It means there were too many of fibers with different diameter.

**Table 6.1.** Shear viscosities of the Nylon 6,6/PEG blend solutions, average diameter and standard deviation of electrospun nanofibers at applied voltage of 15 kV.

| Nylon 6,6/PEG blend ratio | 100/0 (15% solution) | 100/50 | 100/70 | 100/90 | 100/110 | 100/130 | 100/150 |
|---|---|---|---|---|---|---|---|
| Shear viscosity (cP) | 399 | 479 | 419 | 425 | 427 | 590 | 799 |
| Average fiber diameter(nm) | 128 | 178 | 166 | 169 | 180 | 259 | 507 |
| Standard deviation of fiber diameter (nm) | 25 | 40 | 36 | 26 | 20 | 60 | 306 |

Increase in fiber diameter with content of PEG in the blends may be explained by the increase of solution viscosity due to the interaction between the OH groups of PEG and amide groups of nylon 6,6 chains. Table 6.1 shows shear viscosity of the blend nylon 6,6/PEG solutions. It is clear that with increasing the PEG content in the blend solutions, the shear viscosity increases too. Therefore, higher concentrations of PEG in the blend resulted in increased polymer chain entanglements significantly. During electrospinning, the stable jet ejected from Taylor's cone [19] is subjected to tensile stresses and may undergo significant elongational flow. The nature of this elongational flow may determine the degree of stretching of the jet. The characteristics of this elongational flow can be determined by elasticity and viscosity of the solution. The results show that viscosity of the blend solution increases at higher PEG content. Therefore, jet stretching during the electrospinning is less effective at higher PEG content. As a result, the fibers diameters increase with increasing PEG content in the blends. Finally, at the blend ratio of 100/150, the viscosity was too high inhibiting the electrospinning. The solution exit from the needle hardly and could not form a stable jet so that no fiber is collected. Therefore, the shear viscosity must be between suitable ranges to perform the electrospinning process.

## FTIR Characterization

The FTIR spectra of nylon 6,6 and nylon 6,6/PEG composite nanofiber are shown in Fig. 6.4. The peak at 1645 cm$^{-1}$ corresponding to the α-helix amide I band and the peak at 1540 cm$^{-1}$ corresponding to the α-helix amide II band, which exist in both samples. The O-H bending vibration occurs at 1365 cm$^{-1}$ and the peak at 1120 cm$^{-1}$ is assigned to C-O-C stretching. These two last peaks indicate the existence of PEG between the polymer chains of nylon 6,6 and prove that the composite sample in fact is a PCM/ polymer nanofibrous structure.

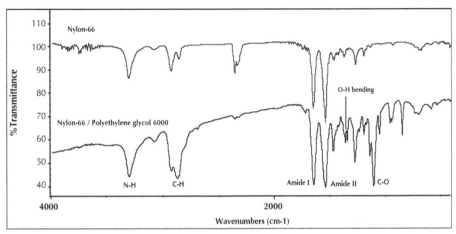

**Figure 6.4.** FTIR spectra of Nylon 6,6 and Nylon 6,6/PEG nanofibers.

## Thermal Properties of Electrospun Nanofibers

In order to analyze the thermal properties of nylon 6,6/PEG nanofiber, pure nylon 6,6 nanofiber mat, and nylon 6,6/PEG nanofiber mats at the blend ratio of 100/50 and 100/130 were chosen. Their DSC thermograms are shown in Fig. 6.5. It is clear that the pure nylon 6,6 nanofiber mat has no phase change peak, no heat capacity, and therefore no ability to heat storage. However, nylon 6,6/PEG samples have melting and crystallization peak. Phase change points and latent heat of phase changes are shown in Table 6.2. It is clear that increasing PEG content in the blend nanofibers has a little effect on the phase change temperatures, but strongly affects the latent heat of phase changes. Figure 6.6 summarizes the theoretical and the experimental values of the latent heat for the blend nanofibers. From the theoretical point of view, the enthalpy values of nylon 6,6/PEG composite nanofibers calculated by multiplying the latent heats of pure PEG and its mass percentage in the composite structure. Figure 6.6 shows that all the experimental values were lower than their corresponding theoretical values. It can be explained by retardation of crystallization process of PEG in the composite structure during electrospinning; because firstly, the cooling of nanofiber and evaporating of solvent during the moving of jet through the air gap did not let the molecular chains of PEG to form a fine crystalline structure in the nanofiber due to being taken place in a little fraction of a second; and secondly, the hydrogen bonding between the hydroxyl group of PEG and carbonyl group of nylon 6,6 also hindered the crystallization process. So, it is concluded that the crystalline areas of PEG in the composite nanofibrous structure are very tiny and encircled by hydrogen bondings. These led to lower crystalline areas of PEG and higher deviation of enthalpies from theoretical values. It is worthy to state that the diluents effect of nylon 6,6 in the composite structure also affects the variation of enthalpy values.

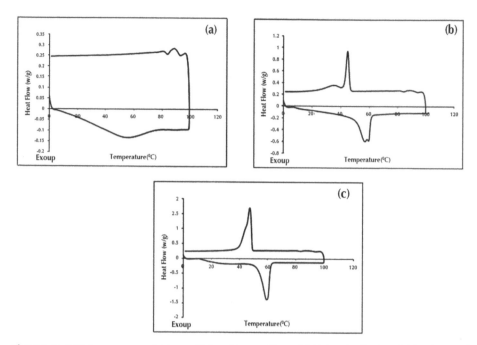

**Figure 6.5.** DSC thermogram of (a) pure Nylon 6,6 nanofibers, (b) Nylon 6,6/PEG with blend ratio of 100/50, and (c) Nylon 6,6/PEG with blend ratio of 100/130.

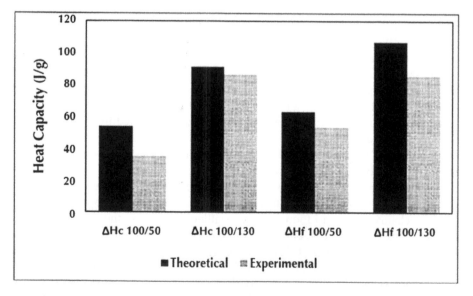

**Figure 6.6.** Theoretical and experimental values of enthalpies for electrospun Nylon 6,6/PEG nanofibers.

Table 6.2. Phase change points and latent heat of Nylon 6,6/PEG nanofibers.

| Blend Ratio | Crystallization Point (°C) | Latent Heat of Crystallization (J/g) | Melting Point (°C) | Latent Heat of Melting (J/g) |
|---|---|---|---|---|
| 100/50 | 44.76 | 34.78 | 57.06 | 53.03 |
| 100/130 | 47.50 | 85.95 | 59.76 | 85.42 |

## CONCLUSION

Nylon 6,6/PEG6000 composite nanofibers have been prepared via electrospinning at the blend ratio of 100/50–100/150. With increasing the total content of PEG in the spinning solution at the constant applied voltage of 15 kV, the average fiber diameter increases too. Increase in the shear viscosity of the blend solutions explains increase in nanofiber diameter with PEG content. With increasing the PEG content up to the blend ratio of 100/130, nanofibers were obtained on the collector. At the blend ratio of 100/150, no defect free fibers could be obtained which was due to the high viscosity of the spinning solution at this blend ratio. In order to investigate the effect of applied voltage on the morphology of the nanofibers, the samples were electrospun at 10 and 20 kV. Increasing the applied voltage led to increase in the average fiber diameter. It can be explained by the fact that increasing the applied voltage draws more solution out of the capillary. Thermal properties of electrospun nanofibers examined with DSC. Pure nylon 6,6 nanofibers showed no heat capacity but the composite nylon 6,6/PEG nanofibers offered the ability of heat storage. It is clear that increasing the PEG content in the blend nanofibers has a little effect on the phase change temperatures, but strongly affects the latent heat of phase changes.

## KEYWORDS

- **Electrospinning**
- **Elongational flow**
- **Phase change materials**
- **Polyethylene glycol**

# Chapter 7

## Recent Advances of Carbon Nanotube/Biopolymers Nanocomposites: A Technical Review

Z. Moridi and V. Mottaghitalab

### INTRODUCTION

Recently, the words "nano-biocomposites" or "biopolymer nanocomposites" are most frequently observed in green environmental research studies. The synthetic polymers have been widely used in a various application of nanocomposites. However, they become a major source of waste after use due to their poor biodegradability. Also, most of the synthetic polymers are not biocompatibe *in vivo* and/or *in vitro* environments. Hence, scientists were interested to biopolymers as biodegradable materials [1]. Later, several groups of natural biopolymers such as polysaccharide, proteins, and nucleic acids were used as a substitute for synthetic polymers in various applications [2]. Nevertheless, the use of these materials has been limited due to relatively poor mechanical properties. Therefore, tremendous efforts have been made to improve the properties of biopolymers as a matrix using of reinforcing agents [3].

Chitosan (CS) is a polysaccharide biopolymer that has been widely used as a matrix in nano-biocomposites. Although, CS represents high biocompatibility and biodegradibility, but biopolymer needs to be mechanically strong using material with superb mechanical properties [4]. Following discovery of carbon nanotube (CNT), results of characterization represented unique electrical and mechanical properties. Thereby, many research studies have focused on improving the physical properties of biopolymer nanocomposites by using of the fundamental behavior of carbon nanotubes [5].

The overall aim of current review is summarizing the recent advances in the production of CNTs/CS nanocomposites.

### BIOPOLYMERS

Biomaterial has been defined as biocompatibility materials with the living systems. The biocompatibility implies the chemical, physical (surface morphology), and biological suitability of an implant surface to the host tissues. S. Ramakrishna et al. reviewed various biomaterials and their application over the last 30 years. They represented applications of biopolymers and their biocomposites in medical applications [6]. These materials can classify to natural and synthetic biopolymers. Synthetic biopolymers have been provided cheaper with high mechanical properties. The low biocompatibility of synthetic biopolymers compared with natural biopolymers such as polysaccharides, lipids, and proteins lead to have paid great attention to the natural biopolymers. On the other hand, the natural biopolymers usually have weak mechanical properties. Therefore, many efforts have been done for improving their properties by blending some filler [7].

Among the natural biopolymers, polysaccharides seem to be the most promising materials in various biomedical fields. These biopolymers have various resource including animal origin, plant origin, algal origin, and microbial origin. Among various polysaccharides, CS is the most usual due to its chemical structure [8].

## Chitosan

Chitin (Fig. 7.1) is the second most abundant natural polymer in the world and extracted from various plant and animals [9]. However, derivations of chitin have been noticed due to insolubility of chitin in aqueous media. CS (Fig. 7.2) is deacetylated derivation of chitin with the form of free amine. Unlike chitin, CS is soluble in diluted acids and organic acids. Polysaccharides are containing 2-acetamido-2-deoxy-β-D-glucose and 2-amino-2-deoxy-β-D-glucose. Deacetylation of chitin converts acetamide groups to amino groups [10]. Degree of deacetylation (DD) is one of the important effective parameters in CS properties and has been defined as "the mole fraction of deacetylated units in the polymer chain" [11].

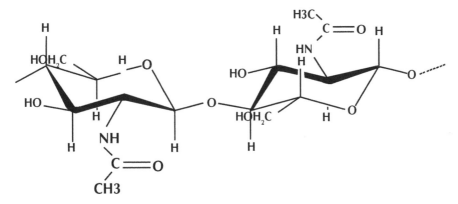

**Figure 7.1.** Structure of chitin.

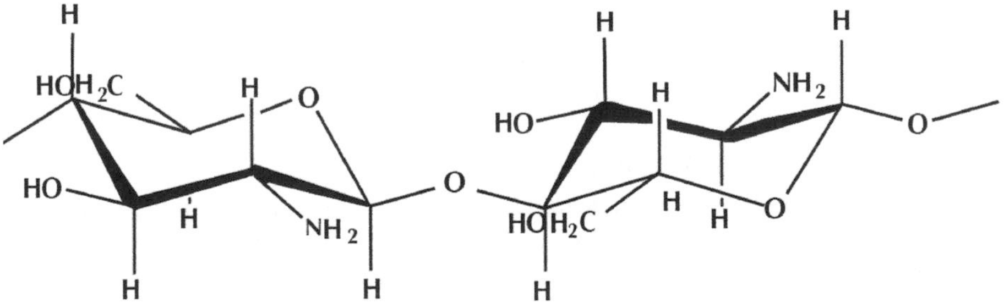

**Figure 7.2.** Structure of chitosan.

The CS could be suitably modified to impart desired properties due to the presence of the amino groups. Hence, a wide variety of applications for CS has been reported over the recent decades. Table 7.1 shows CS applications in variety of fields and their principal characteristics. The supreme biocompatibility [12] and biodegradability [13] of CS yield most potential applications in biomedical [14].

**Table 7.1.** The applications of chitosan in diverse area and their principal characteristics.

| Chitosan Application | | Principal Characteristics | Ref. |
|---|---|---|---|
| Water engineering | | Metal ionic adsorption | [15] |
| biomedical application | Biosensors and immobilization of enzymes and cells | Biocompatibility, biodegradability to harmless products, non-toxicity, antibacterial properties, gel forming properties, and hydro-philicity, remarkable affinity to proteins | [16] |
| | Antimicrobial and wound dressing | Wound healing properties | [17] |
| | Tissue engineering | Biocompatibility, biodegradable, and antimicrobial properties | [18] |
| | Drug and gene delivery | Biodegradable, non-toxicity, biocompatibility, high charge density, mucoadhesion | [19] |
| | Orthopedic/periodontal application | Antibacterial | [20] |
| Photography | | Resistance to abrasion, optical characteristics, film forming ability | [21] |
| Cosmetic application | | Fungicidal and fungi static properties | [22] |
| Food preservative | | Biodegradability, biocompatibility, antimicrobial activity, non-toxicity | [23] |
| Agriculture | | Biodegradability, non-toxicity, antibacterial, cells activator, disease, and insect resistant ability | [24] |
| Textile industry | | Microorganism resistance, absorption of anionic dyes | [25] |
| Paper finishing | | High density of positive charge, non-toxicity, biodegradability, biocompatibility, antimicrobial, and antifungal | [26] |
| Solid-state batteries | | Ionic conductivity | [27] |
| Chromatographic separations | | The presence of free -NH2, primary -OH, secondary -OH | [28] |
| Chitosan gel for LED and NLO applications | | Dye containing chitosan gels | [29] |

## Nano-Biocomposites With Chitosan Matrix

CS due to biocompatibility and biodegradability shows a great potential in biomedical applications. However, the low physical properties of CS are most important challenge that has limited their applications. The development of high performance CS biopolymers has received by incorporating fillers that display a significant mechanical reinforcement [30].

Polymer nanocomposites are reinforced by nano-sized particles with high surface area to volume ratio including nano-particles, nano-platelet, nano-fibers, and CNTs. Nowadays, CNTs have been considered as highly potential fillers for improving of the physical and mechanical properties of biopolymers [31]. Following these reports, researcher assessed the effect of CNTs fillers in CS matrix. Results of these research studies showed appropriate properties of CNTs/CS nano-biocomposites with high potential of biomedical science.

## CARBON NANOTUBES

The CNT, a tubular form of Buckminster fullerene, was discovered by Iijima in 1991 [32]. These are straight segments of tube with arrangements of carbon hexagonal units [33, 34]. Scientists have been greatly attracted to CNTs because of the superior electrical, mechanical, and thermal properties [35]. CNTs can be classified as single walled carbon nanotubes (SWNTs) formed by a single graphene sheet, and multi walled carbon nanotubes (MWNTs) formed by several graphene sheets wrap around the tube core [36]. The typical range of diameters of CNTs are a few nanometers (~0.8–2 nm at SWNTs [37, 38] and ~10–400 nm at MWNTs [39]), and their lengths are up to several micrometers [40].There are three significant methods for synthesizing CNTs including arc-discharge [41], laser ablation [42], and chemical vapor deposition (CVD) [43]. The production of CNTs also can be realized by other synthesis techniques such as, the substrate [44] the sol-gel [45], and gas phase metal catalyst [46].

The C−C covalent bond between the carbons atom are similar to graphite sheets formed by $sp^2$ hybridization. As the result of this structure, CNTs exhibit a high specific surface area (about $10^3$ $m^2/gr$) [47] and thus a high tensile strength (more than 200 GPa) and elastic modulus (typically 1–5 TPa) [48]. CNTs have also very high thermal and electrical conductivity. However, these properties are different according to employed synthesis methods, defects, chirality, the degree of graphitization, and diameter [49]. For instance, the CNT can be metallic or semiconducting, depending on the chirality [50].

Preparation of CNT solution is a challenging area due to strong van der waals interaction between several nanotubes leads led to nanotube aggregation into bundle and ropes [51]. Therefore, the various chemical and physical modification strategies is necessary for improving their chemical affinity [52]. There are two approaches to the surface modification of CNTs including the covalent (grafting) and non-covalent bonding (wrapping) of polymer molecule onto the surface of CNTs [53]. In addition, the reported cytotoxic effects of CNTs *in vitro* may be mitigated by chemical surface modification [54]. On the other hand, studies show that the end-caps on nanotubes are more reactive than sidewalls. Hence, adsorption of polymers onto surface of CNTs can be utilized together with functionalization of defects and associated carbons [55].

The chemical modification of CNTs by covalent bonding is one of the important methods for improving their surface characteristics. Because of the extended π-network of the $sp^2$-hybridized nanotubes, CNTs have a tendency for covalent attachment which introduces the $sp^3$-hibrydized C atoms [56]. These functional groups can be attached to termini of tubes by surface-bound carboxylic acids (grafting to), or direct sidewall modifications of CNTs that are based on the "*in situ* polymerization processing" (grafting from) [57]. Chemical functionalization of CNTs creates various activated groups (such as carboxyl [58], amine [59], fluorine [60], etc.) onto the CNTs surface by covalent bonds. However, there are two disadvantages for these methods. Firstly, the CNT structure may be decomposed due to functionalization reaction [61] and long ultrasonication process [62]. The disruption of $\pi$ electron system is reduced as the result of these damages leading to reduction of electrical and mechanical properties of CNTs. Secondly, the acidic and oxidation treatments which are often used for the functionalization of CNTs are environmentally unfriendly [63]. Thus, non-covalent functionalization of CNTs has been greatly focused because of preserving their intrinsic properties while improving solubility and processability. In this method, non-covalent interaction between the $\pi$ electrons of $sp^2$ hybridized structure at sidewalls of CNTs and other $\pi$ electrons is formed by $\pi$-$\pi$ stacking [64]. These non-covalent interactions can raise between CNTs and amphiphilic molecules (surfactants) (Fig. 7.3a) [65], polymers [66], and biopolymers such as DNA [67], polysaccharides [68], and so forth. In the first method, surfactants including non-ionic surfactants, anionic surfactants, and cationic surfactants are applied for functionalization of CNTs. The hydrophobic parts of surfactants are adsorbed onto the nanotubes surface and hydrophilic parts interact with water [69]. Polymers and biopolymers can functionalize CNTs by using of two methods including endohedral (Fig. 7.3b) and wrapping (Fig. 7.3c). In former method, nanoparticles such as proteins and DNA are entrapped in the inner hollow cylinders of CNTs [70]. But in latter, the van der waals interactions and $\pi$-$\pi$ stacking between CNTs and polymer lead to the wrapping of polymer around the CNTs [71]. Various polymers and biopolymers such as polyaniline [72], DNA [73], and CS [74] interact physically through wrapping of nanotube surface and $\pi$-$\pi$ stacking by solubilized polymeric chain. However, Jian et al. (2002) developed a technique for the non-covalent functionalization of SWNTs most similar to $\pi$-$\pi$ stacking by PPE without polymer wrapping [75].

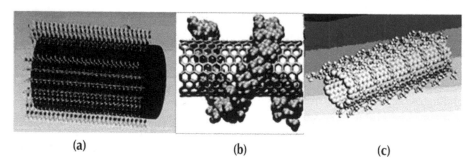

(a)                    (b)                    (c)

**Figure 7.3.** Non-covalent functionalization of CNTs by (a) surfactants, (b) wrapping, (c) endohedral.

These functionalization methods can provide many applications of CNTs. In this context, one of the most important applications of CNTs is biomedical science such as biosensors [76], drug delivery [77], and tissue engineering [78].

## Nanotube Composites

Regardless of biological properties, the electrical, mechanical, and thermal properties of biopolymers need to be reinforced using suitable filler for diverse applications. Following discovery of CNTs, their usage as filler in polymer matrix improves the bulk properties compared to neat matrix [79]. Ajayan et al. in 1994 introduced CNTs as filler in epoxy resin through the alignment method [80]. Later, many studies have focused on CNTs as excellent substitute for conventional nanofillers in the nanocomposites. Recently, many polymers and biopolymers have been reinforced by CNTs. As mentioned earlier, these nanocomposites have remarkable characteristics compared with bulk materials due to their unique properties [81].

There are several parameters affect the mechanical properties of composites including proper dispersion, large aspect ratio of filler, interfacial stress transfer, well alignment of reinforcement, and solvent choice [82].

Uniformity and stability of nanotube dispersion in polymer matrix are most important parameters for evaluation of composite performance. In fact the prefect filler distribution is a prerequisite for efficient load transfer from matrix to filler [83]. The correlation between effective dispersion and functionalization and their effects on the properties of CNT/polymer nano-composites were extensively investigated. In overall, it has been showed that the proper dispersion enhances a variety of mechanical properties of nanocomposites [71].

Fiber aspect ratio, defined as "the ratio of average fiber length to fiber diameter." This parameter is one of the main effective parameters on the longitudinal modulus [84]. CNTs generally have high aspect ratio but their ultimate performance in a polymer composite is different. The high aspect ratio of dispersed CNTs could lead to a significant load transfer [85]. However, the aggregation of the nanotubes decreases the effective aspect ratio of the CNTs. This is one of the processing challenges regarding to poor CNTs dispersion [86].

The interfacial stress transfer is essential for load transfer process while external stresses apply to the composites. Experimental observation showed that fillers take a significant larger share of the load due to CNTs-polymer matrix interaction. Also, the mechanical properties of polymer nanotube composites represent an enhancement in Young's modulus due to adding CNTs [87]. Wagner et al. investigated the effect of stress-induced fragmentation of MWNTs in a polymer matrix. The results showed that generated tensile through polymer deformation can be thoroughly transferred to CNTs [88].

The CNT alignment in polymer matrix is extremely effective parameter to explain the properties of CNT composites. Quin Wang et al. [89], for instance, assessed the effects of CNT alignment on electrical conductivity and mechanical properties of SWNT/epoxy nanocomposites. The electrical conductivity, Young's modulus, and tensile strength of the SWNT/epoxy composite rise with increasing SWNT alignment due to increase of interface bonding of CNTs in the polymer matrix.

Umar khan et al. in 2007 examined the effect of solvent choice on the mechanical properties of CNTs-polymer composites. They were fabricated double-walled nanotubes (DWNT) and polyvinyl alcohol (PVA) composites into the different solvents including water, DMSO, and NMP. This work shows that solvent choice can have a dramatic effect on the mechanical properties of CNTs-polymer composites [90]. Also, a critical CNTs concentration has defined as optimum improvement of mechanical properties of nanotube composites where a fine network of filler formed [91]. There are other effective parameters in mechanical properties of nanotube composite such as size, crystallinity, crystalline orientation, purity, entanglement, and straightness. Generally, the ideal CNT properties depend on matrix and application [92].The various functional groups on CNT surface enable to couple with polymer matrix. Strong interfacial interaction creates efficient stress transfer. As previously pointed out, stress transfer is a critical parameter to control the mechanical properties of composite. However, covalent treatment of CNT reduce electrical [93], and thermal [94, 95] properties of CNTs. These reductions affect on ultimate properties of CNTs.

Polymeric matrix may wrap around CNT surface by non-covalent functionalization. This process causes improvement in composite properties through various specific interactions [96]. In this context, Gojny et al. [97] evaluated electrical and thermal conductivity in CNTs/epoxy composites. Figures 7.4 and 7.5 show respectively electrical and thermal conductivity in various filler content including carbon black (CB), SWNT, DWNT, and functionalized MWNT. The experimental results represented that the electrical and thermal conductivity in nanocomposites improve by non-covalent functionalization of CNTs.

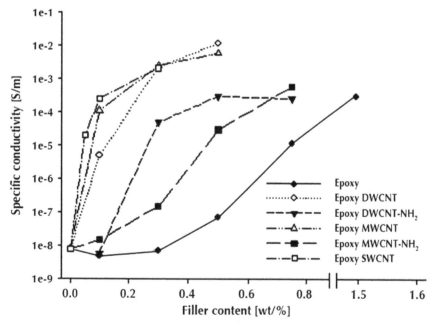

**Figure 7.4.** Electrical conductivity of the nanocomposites as function of filler content in weight percent.

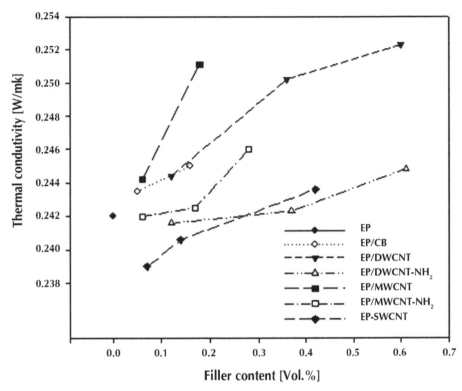

**Figure 7.5.** Thermal conductivity as function of the relative provided interfacial area per gram composite (m2/g).

## Mechanical and Electrical Properties of CNT/Natural Biopolymer Composites

Table 7.2 represents mechanical and electrical information of CNTs/natural polymer compared with neat natural polymer. These investigations show the higher mechanical and electrical properties of CNTs/natural polymers than neat natural polymers.

**Table 7.2.** Mechanical and electrical information of neat biopolymers compared with their carbon nanotube nanocomposites.

| Method | Biopolymer | Mechanical | | | Conductivity | Ref. |
|---|---|---|---|---|---|---|
| | | Elastic modulus (Mpa) | Tensile strength (Mpa) | Strain to failure (%) | | |
| Polymerized hydrogel | Neat collagen | | NA | | 11.37ms ± 0.16 | [98] |
| | collagen/CNTs | | | NA | 11.85ms ± 0.67 | |
| Solution-evaporation | Neat chitosan | 1.08 ± 0.04 | 37.7 ± 4.5 | | 0.021 nS/cm | [99, 100] |
| | chitosan/CNTs | 2.15 ± 0.09 | 74.3 ± 4.6 | NA | 120 nS/cm | |

**Table 7.2.** *(Continued)*

| Method | Biopolymer | Elastic modulus (Mpa) | Tensile strength (Mpa) | Strain to failure (%) | Conductivity | Ref. |
|---|---|---|---|---|---|---|
| Wet-spinning | Neat chitosan | 4250 | NA | NA | | [101] |
| | chitosan/CNTs | 10,250 | | | | |
| Electrospinning | Neat silk | 140 ± 2.21 | 6.18 ± 0.3 | 5.78 ± 0.65 | 0.028 S/cm | [102] |
| | silk/CNTs | 4817.24 ± 69.23 | 44.46 ± 2.1 | 1.22 ± 0.14 | 0.144 S/cm | |
| Dry-jet wet-spinning | Neat cellulose | 13,100 ± 1100 | 198 ± 25 | 2.8 ± 0.7 | Negligible | [103, 104] |
| | cellulose/CNTs | 14,900 ± 13 00 | 257 ± 9 | 5.8 ± 1.0 | 3000 S/cm | |
| Electrospinning | Neat cellulose | 553 ± 39 | 21.9 ± 1.8 | 8.04 ± 0.27 | | [105] |
| | cellulose/CNT | 1144 ± 37 | 40.7 ± 2.7 | 10.46 ± 0.33 | | |

## Carbon Nanotube Composite Application

Great attention has been paid in recent years to applying nanotube composites in various fields. Wang and Yeow [106] reviewed nanotubes composites based on gas sensors. These sensors play important role for industry, environmental monitoring, and biomedicine. The unique geometry, morphology, and material properties of CNTs led to apply them in gas sensors. There are many topical studies for biological and biomedical applications of CNT composites due to its biocompatibility [107]. These components promoted biosensors [108], tissue engineering [95], and drug delivery [109] fields in biomedical technology. On the other hand, light weight, mechanical strength, electrical conductivity, and flexibility are significant properties of CNTs for aerospace applications [110].

Kang et al. [111] represented an overview of CNT composite applications including electrochemical actuation, strain sensors, power harvesting, and bioelectronics sensors. They presented appropriate elastic and electrical properties for using nanoscale smart materials to synthesis intelligent electronic structures. In this context, Mottaghitalab and coworkers developed polyaniline/SWNTs composite fiber [112] and showed high strength, robustness, good conductivity, and pronounced electroactivity of the composite. They presented new battery materials [113] and enhancement of performance artificial muscles [114] by using of these CNT composites.

Thai Ong et al. [115] addressed sustainable environment and green technologies perspective for CNT applications. These contexts are including many engineering fields such as wastewater treatment, air pollution monitoring, biotechnologies, renewable energy technologies, and green nanocomposites.

Sariciftci et al. [116] first time discovered photo induced electron transfer from CNTs. Later, optical and photovoltaic properties of CNT composites have been studied by many groups. Results suggested the possible creation of photovoltaic devices due to hole-collecting electrode of CNTs [117].

Food packaging is another remarkable application of CNT composites. Usually, poor mechanical and barrier properties have limited applying biopolymers. Hence, appropriate filler is necessary for promotion of matrix properties. Unique properties of CNTs has been improved thermal stability, strength and modulus, and better water vapor transmission rate of applied composites in this industrial [118].

## CHITOSAN/CARBON NANOTUBE COMPOSITES

In recent decade, scientists interested to creation of CS/CNTs composite due to providing unexampled properties of this composite. They attempted to create new properties by adding the CNTs to CS biopolymers. At recent years, several research articles were published in variety of applications. Figure 7.6 summarizes the application of CS/CNTs nanocomposies.

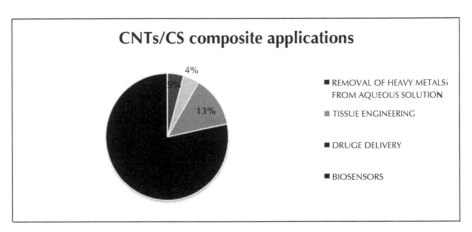

**Figure 7.6.** The graph of CNTs/CS nanocomposite application.

### Chitosan/Carbon Nanotube Nanofluids

Viscosity and thermal conductivity of nanofluids containing MWNTs stabilized by CS were investigated by Phuoc et al. [119]. It has been shown that thermal conductivity enhances significantly higher than the predicted amount using the Maxwell's theory. In addition, they observed that dispersing CS into deionized water increased the viscosity of nanofluids significantly and its behavior switches to non-Newtonian fluid.

### Preparation Methods of CNTs/CS Nanocomposites

There are several methods for creation of nano-biocomposites. Among them, researchers have studied some of these methods for preparation of the CNTs/CS nanocomposites.

#### Solution-casting-evaporation

Electrochemical sensing of CNT/CS in a media containing dehydrogenase enzymes for preparing glucose biosensor initially investigated in 2004 [120]. The nanotube composite has been prepared using of solution-casting-evaporation method. In this method, the CNT/CS films were prepared by casting and then drying of CNT/CS solution on the surface of glassy carbon electrode. This CNT/CS system showed a new biocomposite platform for development of dehydrogenase-based electrochemical biosensors due to providing a signal transducing of CNT. The great results of this composite in biomedical application led to many studies in this context.

The effect of CNT/CS matrix on direct electron transfer of glucose oxidase (GOD) and glucose biosensor was examined by Liu and Dong et al. [121]. They exhibited high sensitivity and better stability of CNT/CS composites compared with pure CS films. Furthermore, Tkac et al. [122] used the SWNT/CS films for preparation a new galactose biosensor with highly reliable detection of galactose. Tsai et al. [123] immobilized lactate dehydrogenase within MWNT/CS nanocomposite for producing lactate biosensors. This proposed biosensor provided a fast response time and high sensitivity. Also, Zhou and chen et al. [124] showed the immobilization of GOD molecules into CS wrapped SWNT film is an efficient method for the development of a new class of very sensitive, stable, and reproducible electrochemical biosensors.

Several experiments were performed on DNA biosensor based on CS film doped with CNTs by Yao et al. [125]. They found that CNT/CS film can be used as a stable and sensitive platform for DNA detection. The results demonstrated to improve sensor performance by adding CNT to CS film. Moreover, the analytical performance of glassy carbon electrodes modified with a dispersion of MWNT/CS for quantification of DNA was reported by Bollo et al. [126]. The new platform immobilizes the DNA and opens the doors to new strategies for development of biosensors. In other experiments, Zeng et al. [127] reported high sensitivity of glassy carbon electrode modified by MWNT-CS for cathodic stripping voltammetric measurement of bromide (Br). Qian et al. [128] prepared amperometric hydrogen peroxide biosensor based on composite film of MWNT/CS. The results showed excellent electrocatalytical activity of the biosensor for $H_2O_2$ with good repeatability and stability.

Liu and Dong et al. [129] reported effect of CNT/CS matrix on amperometric lactase biosensor. Results showed some major advantages of this biosensor involving detecting different substrates, possessing high affinity and sensitivity, durable long-term stability, and facile preparation procedure.

Gordon Wallace and his coworkers [130] with particular attention paid to preparing of SWNT/CS film by solution-cast method and then characterized their drug delivery properties. They found that the SWNT/CS film decrease the release rate of dexamethasone. Growth of apatite on CS-MWNT composite membranes at low MWNT concentrations was reported by Yang et al. [131]. Apatite was formed on the composites with low concentrations. Immune-sensors can detect various substances from bacteria to environmental pollutants. CNT/CS nano-biocomposite for immune-sensor fabricated by kaushik et al. [132]. Electron transport in this nano-biocomposite enhanced and improved the detection of ochratoxin-A, due to high electrochemical properties of SWNT. Also, CNT/CS nanocomposite used for detection of human chorionic gonadotrophin antibody was performed by Yang et al. [133] and displayed high sensitivity including good reproducibility.

*Properties and characterization*

Wang et al. [99] represented that morphology and mechanical properties of CS has promoted by adding CNTs. Beside, Zheng et al. [134] proved that conducting direct electron is very useful for adsorption of hemoglobin in CNT/CS composite film. These studies have been demonstrated that this nano-biocomposite can used in many field such as biosensing and biofuel cell approaches.

Tang et al. [135] evaluated water transport behavior of CS porous membranes containing MWNTs. They characterized two nanotube composites with low molecular weight CSP6K and high molecular weight CSP10K. Because of hollow nano-channel of MWNTs located among the pore network of CS membrane, the water transport results for CSP6K enhanced, when the MWNTs content is over a critical content. But, for CSP10K series membranes, the water transport rate decreased with increase of MWNTs content due to the strong compatibility effect of MWNTs.

In another attempts, the CNT/CS nanocomposites were prepared by utilizing poly(styrene sulfonic acid)-modified CNTs [136]. Thermal, mechanical, and electrical properties of CNT/CS composite film prepared by solution-casting showed potential applications for membranes and sensor electrodes.

*Novel approaches*

In a new approach, MWNT functionalized with—COOH groups at the end or at the sidewall defects of nanotubes by CNTs in nitric acid solvent. The functionalized CNTs immobilized into CS films by Emilian Ghica et al. [137]. This film applied in amperometric enzyme biosensors, resulted glucose detection, and high sensitivity.

In a novel method, Kandimalla and Ju [138] cross-linked CS with free—CHO groups by glutaraldehyde and then MWNTs added to the mixture. The cross-linked MWNT-CS composite immobilized acetylcholinesterase (AChE) for detecting of both acetylthiocholine and organophosphorous insecticides. On the other hand, Du et al. [139] created a new method for cross linking of CS with carboxylated CNT. This new method was performed by adding glutaraldehyde to MWNT/CS solution. They immobilized AChE on the composite for preparing an amperometric acetylthiocholine sensor. The suitable reproducible fabrication, rapid response, high sensitivity, and stability could provide an amperometric detection of carbaryl and treazophos [140] pesticide. Abdel Salam et al. [141] showed the removal of heavy metals including copper, zinc, cadmium, and nickel ions from aqueous solution in MWNT/CS nanocomposite film.

*Surface deposition crosslink*

Liu et al. [142] decorated CNT with CS by surface deposition and cross linking process. In this new method, CS macromolecules as polymer cationic surfactants were adsorbed on the surface of the CNTs. CS is able to produce stable dispersion of the CNT in acidic aqueous solution. The pH value of the system was increased by ammonia solution to become non-dissolvable of CS in aqueous media. Consequently, the soluble CS deposited on the surface of CNTs similar to CS coating. Finally, the surface-deposited CS was cross-linked to the CNTs by glutaraldehyde. They found potential applications in biosensing, gene, and drug delivering for this composite.

*Electro deposition method*

Luo and Chen et al. [143] used nanocomposite film of CNT/CS as glucose biosensor by a simple and controllable method. One-step electro deposition method comprise a pair of gold electrodes connects to a power supply and then dipped into the CNT/CS solution. The pH near the cathode surface increases when solubility of CS decreases. However, CS becomes insoluble and entrapped CNT can be deposited onto the cathode surface at pH about 6.3.

Yao et al. [144] also characterized electro catalytic oxidation and sensitive electro analysis of NADH on a novel film of CS-DA-MWNTs and improved detection sensitivity. In this new method, glutaraldehyde cross-linked CS-DA with the covalent attachment of DA molecules to CS chains. Following, solution of MWNT dispersed in CS-DA solution dropped on an Au electrode for preparing CS-DA-MWNTs film and finally dried.

*Covalently grafting*

Carboxylic acid (–COOH) groups can be covalently grafted on CNTs sidewalls by refluxing of CNTs in acidic solution. The carboxylated CNTs then adds to aqueous solution of CS. Grafting reactions were accomplished by $N_2$ purging, and then heating to 98°C of CNTs/CS solution. Shieh et al. [145] compared mechanical properties of CNTs-grafted-CS to the ungrafted CNTs. Significant improvement storage modulus and water stability of the CS nanocomposites.

Wu et al. [146] created another process for make a CS-grafted MWNT composite. In this different method, after preparing oxidized MWNT (MWNT-COOH), they generate the acyl chloride functionalized MWNT (MWNT-COCl) in a solution of thionyl chloride. In the end, the MWNT-grafted-CS was synthesized by adding CS to MWNT-COCl suspension in anhydrous dimethyl formamide. The covalent modification has improved interfacial bonding and resulted high stability of CNT dispersion. Biosensors and other biological applications are evaluated as potential usage of this component. Also, Carson et al. [147] prepared a similar composite by reacting CNT-COCl and CS with potassium persulfate, lactic acid, and acetic acid solution at 75°C. They estimated that the CNT-grafted-CS composite can be used in bone tissue engineering because the improvement of thermal properties.

*Nucleophilic substitution reaction*

Covalent modification of MWNT was accomplished with a low molecular weight CS (LMCS) by Ke et al. [148]. In this method, the acyl chloride functionalized grafted to LMCS in DMF/Pyridine solution. This novel derivation of MWNTs can be solved in DMF, DMAc, and DMSO, but also in aqueous acetic acid solution.

*Electrostatic interaction*

Furthermore, Baek et al. [149] synthesized CS nanoparticles-coated MWNTs composite by electrostatic interactions between CS particles and functionalized CNT. They prepared CS nanoparticles and CS microspheres by precipitation method and cross linking method respectively. The electrostatic interactions between CS particles solution in distilled deionized water and the carboxylated CNTs were confirmed by changing the pH solution. Results showed same surface charges in pH 2 (both were positively charged) and pH 8 (both were negatively charged). The electrostatic interactions can be caused at pH 5.5 due to different charges between CS particles and CNT with positive and negative surface charges respectively. These CS particles/CNT composite materials could be utilized for potential biomedical.

Also, Zhao et al. [150] constructed SWNTs/phosphotungstic acid modified SWNTs/CS composites using phosphotungstic acid as an anchor reagent to modify SWNTs. They succeed to use $PW_{12}$-modified SWNT with a negative surface charge, and on the contrary, positively charged CS by electrostatic interaction. These strong interfacial interaction between SWNTs and CS matrix presented favorable cytocompatibility for the potential use as scaffolds for bon tissue engineering.

### Microwave irradiation

Yu et al. [151] created a new technique for synthesis of CS modified CNT by using microwave irradiation. In this technique, MWNTs/nitric acid solution treats by microwave radiation and then dries for purification of MWNTs. A mixture of purified MWNTs and CS solution was reacted in the microwave oven and then centrifuged. The pH of black-colored solution adjusted and centrifuged for precipitation of CNT/CS composite. This technique is highly efficient than conventional methods.

### Layer-by-layer

Wang et al. [152] characterized MWNT/CS composite rods with layer-by-layer structure prepared by *in situ* precipitation method. Samples prepared by coating of CS solution on internal surface of a cylindrical tube and then filling with MWNT/CS solution in acetic acid. They examined morphology, mechanical, and thermal properties of this composite rod. The excellent mechanical property of these new composite rods has made potential of bone fracture internal fixation application.

### Layer-by-layer self assembly

Xiao-bo et al. [153] produced a homogeneous multilayer film of MWNT/CS by using layer-by-layer self assembly method. In this method, negatively charged substrates were dipped into poly (ethyleneimine) aqueous solution, MWNTs suspension, and CS solution respectively and dried at the end. In this process both of CS and PEI solution were contained NaCl for the LBL assembly. The films showed stable optical properties and were be appropriate for biosensors applications.

### Freeze-drying

Lau et al. [154] synthesized and characterized a highly conductive, porous, and biocompatible MWNT/CS biocomposite film by freeze-drying technique. This process was performed by freezing and drying of MWNT/CS dispersion into an aluminum mold. Such a composite permitted delivery of needed antibiotics with effect of increased antibiotic efficacy in a patent by Jennings et al. [155].

### Wet-spinning

Gordon Wallace and coworkers [156] recently reported that CS is a good dispersing agent for SWNT. They also demonstrated several methods in preparing SWNT/CS macroscopic structure in the form of films, hydrogels, and fibers [157]. The CNT/CS dispersions in acetic acid were spun into an ethanol: NaOH coagulation solution bath. They were demonstrated increasing mechanical properties of wet spun fibers by improving dispersion [158].

### Electrospinning

In our recent work, the CS/MWNTs composite nanofiber were fabricated using electrospinning. In our experimental researches, different solvents including acetic acid 1–90%, formic acid, and TFA/DCM were tested for the electrospinning of CS/CNT. No jet was seen upon applying the high voltage even above 25 kV by using of acetic acid 1–30% and formic acid as the solvent for CS/CNT. When the acetic acid 30–90%, used as the solvent, beads were deposited on the collector. Therefore, under these conditions, nanofibers were not formed.

The TFA/DCM (70:30) solvent was only solvent that resulted for electrospinnability of CS/CNT. The scanning electron microscopic (Fig. 7.7 showed the homogenous fibers with an average diameter of 455 nm (306–672) were prepared with CS/CNT dispersion in TFA/DCM 70:30. These nanofibers have a potential for biomedical applications.

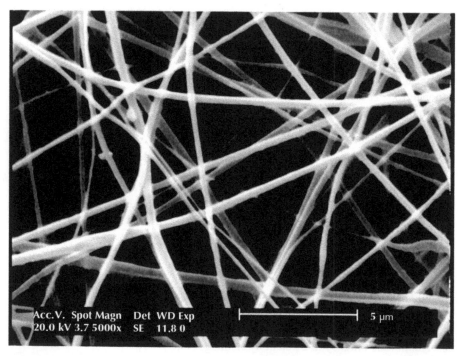

**Figure 7.7.** SEM micrograghs of electrospun fibers of CS/MWNT at chitosan concentration of 10 wt%, 24 kV, 5 cm, TFA/DCM: 70/30 (0.06% wt MWNT).

## CONCLUSION

With less than 10 years history, several 10-research studies have been created in CS biocomposites reinforcement using CNTs. In conclusion, much progress has been made in preparation and characterization of the CNTs/CS nanocomposites. We reported several methods for preparing these nano-biocomposites. In addition, the CNTs/CS applications have been classified including biomedicine (tissue engineering, biosensors, and drug delivery) and wastewater in this review.

Most importantly, the overriding results of electrospinning of CNTs/CS nanocomposites in our recent paper have been discussed. It is expected a high potential application in tissue engineering and drug delivery by these nano-biocomposites. It is believed that, with more attentions to the preparation methods of CNTs/CS nanocomposites and their characterization have a promising future in biomedicine science.

## KEYWORDS

- **Carbon nanotubes**
- **Chitosan**
- **Degree of deacetylation**
- **Polymer nanocomposites**
- **Synthetic biopolymers**

# Chapter 8

## Polypyrrole Coated Polyacrilonitril Electrospun Nanofibers

Hamideh Mirbaha and Mahdi Nouri

### INTRODUCTION

Application of electrically conductive polymers to nanofiber mats is considered. Different concentration of Polyacrylonitrile (PAN) solutions was electrospun. The electrospun PAN nanofibers were coated with polypyrrole (PPy) as conductive polymers. Mean nanofibers diameter and their morphology were studied using scanning electron microscope (SEM). Electrical resistance of the coated mats was measured. The results of the measurements indicated that PPy coating were effective in the reduction of the nanofibers electrical resistance comparing with that of uncoated nanofibers.

Polymers that exhibit high electrical conductivity have successfully been synthesized in the last few decades. These electrically conductive polymers have increasing number of applications in different areas of microelectronics and chemical analysis [1].

Conductive polymers have л-conjugated structure, which can conduct electricity in doped form. Their conductivity can be varied from isolated regime to metallic regime. Their intrinsic electrical conductance together with the properties of conventional polymers has made them good candidates for use in different parts of microelectronics. PPy has been one of the most extensively studied conductive polymers during the past two decades due to its excellent controllable electrical conductivity, environmental stability. As it is reported in several studies one of the nanostructures that can be used in microelectronic devices is nanofiber. Nanofibers are new generation of fibers that have the advantage of high surface to volume ratio due to their tiny size. PPy nanofibers are prepared either solely or by blending with other conventional polymers for use as conductive scaffolds in tissue engineering [2].

The main problems on application of conducting polymers have been brittleness, insolubility, and unstable electrical properties. Due to excellent flexibility of textile materials, chemical oxidative deposition of conducting polymers onto textile surface yields new composite material with potential use in various applications [3]. The resulting composite possesses both the mechanical properties of the textile and the electrical properties of the conducting polymers which can be used as flexible smart materials for application in biomedical and tissue engineering. Rossi et al. [4] reported a sensitized glove based on the PPy coated lycra/cotton fabric. Electrospinning is the simplest, cheapest, and the most straightforward way to produce nanofibers by forcing a polymer melt or solution through a spinneret with an electric field. It is of indispensable importance for the scientific and economical revival of many parts of the developing world. It is an effective method to manufacture ultra fine fibers or fibrous structures of many polymers with diameter in the range from several micrometers down to tens of nanometers [5]. From a polymer solution a charged jet is created, when the electrical force overcomes surface tension. The jet typically develops a bending instability and then solidifies to form fibers, which measures in the range of nanometers to 1 μm.

In this experimental study preparation of electrically conductive PAN nanofiber using insitu chemical polymerization of pyrrole on to the surface of the nanofibers were studied.

## EXPERIMENTAL AND METHODS

### Materials
Polyacrylonitrile was from POLYACRYL IRAN CO. Reagent-grade pyrrole was from Aldrich and was distilled prior to the use. All the other reagents were laboratory grades and used as received.

### Electrospinning
In the electrospinning process, a high electric potential (Gamma High voltage) was applied to a droplet of PAN solution at the tip (0.3 mm inner diameter) of a syringe needle. The electrospun nanofibers were collected on a target plate, which was placed at a distance of 10 cm from the syringe tip. A high voltage in the range from 10 kV to 20 kV was applied to the droplet of solution at the tip.

### Coating of Nanofibers
For the deposition of PPy on the surface of nanofibers, a mat of nanofibers was added to temperature-controlled flask containing distilled water and pyrrole (typically, 20 cc water, 0.01 g pyrrole). The bath was equipped with a magnetic stirrer and polymerization was performed at the room temperature. The webs (about 0.01 g, each) were weighted with an accuracy of 0.001 g. After the stabilization of the temperature, in a period of 10 min, solution of FeCl3 (typically 0.1 g) in water (typically 5 cc) as oxidizing agent was added gradually to the slowly stirring bath. After the required time of reaction (2 h), the samples were removed and rinsed thoroughly. To remove the unreacted chemicals, the samples were rinsed thoroughly with excess of water and methanol and dried at room temperature.

### Characterization
Fiber formation and morphology of the coated nanofibers were determined using a SEM Philips XL-30. A small section of the web was placed on SEM sample holder and coated with gold (BAL-TEC SCD 005 sputter coater). Electrical conductivity of the coated mats was determined employing the standard four-probe technique.

## RESULTS AND DISCUSSION

A series of mats were prepared when the PAN concentration was varied from 8 to 14% at the 15 KV constant electric field. At concentrations below 8% the electrospinning process generated a mixture of fibers and droplet at 15 KV. Electrospinning of solution with concentration higher than 15% was prohibited by their high viscosity and electrospinning was not possible. Figure 8.1 shows typical SEM photomicrographs of the resulted nanofibers from solution of 10% PAN concentration.

It is shown that there is a significant increase in average fiber diameter from 80 to 400 nm with increase in PAN concentration, which shows the important role of the concentration in fiber formation during electrospinning process. One of the most important parameters influencing the fiber diameter is solution viscosity. A higher viscosity results in a larger fiber diameter. The polymer solution viscosity is proportional to polymer concentration and polymer molecular weight. Thus, at fixed polymer molecular weight, the higher the polymer concentration the larger the resulting nanofibers diameter will be. In fact, the fiber diameter will increase with the polymer concentration.

Concentration of the polymer solution reflects the number of entanglements of polymer chains in the solution, thus solution viscosity. Experimental observations in electrospinning confirm that for fiber formation to occur, a minimum polymer concentration is required. Below this critical concentration, application of electric field to a polymer solution results electrospraying and formation of droplets due to the instability of the ejected jet. As the polymer concentration increased, a mixture of beads and fibers are formed.

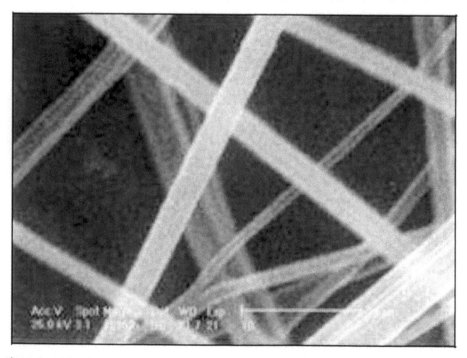

**Figure 8.1.** SEM photomicrograph of electrospun PAN nanofiber at 10% concentration.

Figure 8.2 shows SEM photomicrograph of PPy coated electrospun nanofiber. A uniform layer of PPy is deposited on the nanofibers surfaces. Average fiber diameter for uncoated nanofibers, electrospun from 10 and 15% concentration, was 230 nm and 390 nm, respectively. After coating process average fiber diameter increases to 350 and 500 nm. It means that a layer of PPy with the thickness of approximately 120 nm was coated onto the nanofibers. Measurement of the electrical conductivity of the coated mats shows electrical conductivity in the range of $10^{-3}$ to 1 S/cm depending on the chemical coating conditions such as amount of the pyrrole in the coating process and the mole ratio of the FeCl3 to the used pyrrole.

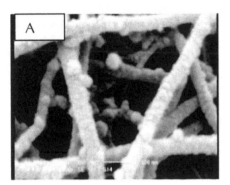

**Figure 8.2.** SEM photomicrograph of PPy coated electrospun PAN nanofiber at 15% (A) and 10% (B) concentration.

## CONCLUSION

Possibility of the chemical deposition of PPy on the surface of the electrospun PAN nanofibers is shown. Diameter of the nanofibers is governed by the concentration of the electrospun PAN solution. A thin layer with the thickness of 120 nm is formed onto the nanofibers surface. The electrical conductivity of the coated nanofiber mats depends on the coating conditions and reaches to 1 S/cm.

## KEYWORDS

- **Coating**
- **Electrospinning**
- **Nanofiber**
- **Polypyrrole**

# Chapter 9

## Semi-empirical AM-1 Studies on Porphyrin

Nazmul Islam and Minakshi Das

---

### INTRODUCTION

The study of molecular interactions has been a great challenge from the experimental as well as theoretical point of view. A lot of endeavors have been made to explain the nature of bonding and reactivity of molecular systems based on some insightful ideas and pragmatic rules.

Porphyrins are a group of biologically active heterocyclic organic pigment of both neutral—most famously as the pigment in red blood cells (RBC), and synthetic origin. The parent porphyrin is porphine, and substituted porphines are called porphyrins. The structural properties of porphyrins are closely related to the important role of these compounds in the biological systems. The porphyrins are usually described as a union of four modified pyrrole subunits interconnected at their a-C atoms via the methine bridges to form a macro cycle. The macro cycle is referred to cyclic molecules large enough to accommodate a metal ion in the interior. The macro cycle has 26 p electrons. As the porphyrins posses (4n + 2) p electrons that are delocalized over the macrocycle, they certainly satisfy the Hückel's rule for [1] and hence are aromatic in nature. The macrocycles are highly-conjugated system, thus are deeply colored. Owing to their high extinction coefficients and tunable fluorescence emission by changing the central metal ion, porphyrin supramolecular complexes are used as good fluorescent pH sensors [2]. The interesting structures of naturally occurring porphyrins, its isomer, and substituted analogs have been perfected by nature to give functional dyes par excellence [3]. The important roles these tetrapyrrolic macrocycles play in vital biological processes namely, in the photosynthesis the Mg-porphyrins complex, chlorophyll plays an important role, in blood, oxygen transport occurs by Fe-porphyrins complexes known as hemoglobin and myoglobin, and in the electron transport process another group of Fe-porphyrins complexes, cytochromes play a vital role. Supramolecules containing zinc-porphyrins have been widely used in host–guest chemistry [4]. The porphyrins are called "Pigments of Life" [5] for their manifold biological significance. Due to the presence of conjugated $\pi$-electrons, porphyrins are highly stable and thus also they are useful in material science as the components in organic metals, molecular wires, and other devices. Moreover, due to the advent of photodynamic therapy, porphyrins and their metallocomplexes are used in the treatment of cancer and dermatological diseases [6]. The porphyrias constitute a heterogeneous group of diseases, all of which exhibit increased excretion of porphyrin or porphyrin precursors. Some form of porphyrias are inherited, where as others are acquired. Recently the anticancer drugs are synthesized using several gold-porphyrin complexes [7]. Thus the interdisciplinary interests on the porphyrins generate a new science relating to the development of novel porphyrin-like molecules. The molecules are designed by the structural variation of the tetrapyrrolic macrocycle while maintaining a (4n + 2) $\pi$ main conjugation pathway anticipated to exhibit special properties.

The probable reasons for the importance of porphyrin complexes in a variety of biological systems are:

1.  They are biologically handy compounds.
2.  Their functions can be varied by changing the metal, its oxidation state, or the nature of the organic substituents on the porphyrin structure.
3.  It is a general principle that the evaluation tends to proceed by modification structures and functions that are already present on the organism rather than producing new ones de novo [8]. As evidenced by the host of expanded, re-shuffled, inverted, contracted, and otherwise modified porphyrins brought to light in recent years, the quest for this concept has proven to be highly successful.

Naturally occurring porphyrins are synthesized by living matter. In laboratory, the porphyrin skelet can be synthesized by several routes based on condensation reactions between aldehydes, pyrroles, dipyrrylmethanes or similar precursors under acidic conditions and following oxidation [9]. Rothmund [10] was the first one who synthesized a porphyrin isomer, tetraphenylporphyrin (TPP) using benzaldehyde and pyrrole. Literature [6] shows that since that time, a series of both symmetrical and asymmetrical porphyrins has been prepared. Having highly symmetrical (up to $D_{4h}$) and conjugated structure porphyrins have very rich electronic absorption properties. This is one of the important reasons which strongly attract quantum chemists to pay attention to them. Semi-empirical calculations resulted in important insights into the electronic structure and spectra of porphyrins [11, 12]. During the development of the quantum chemical method, many of the empirical chemical concepts were derived rigorously and it has provided a method for the calculation of the properties of chemical systems and the bonding that is involved in the formation of molecular systems. While such calculations continue to be useful, especially for studies of porphyrins excited states, recent advances in quantum chemical and computer technologies permit more accurate ab initio calculations [13] on large porphyrin-sized molecules. However, most porphyrin chemists do not consider ab initio calculations as a practical tool in their research [13].

Since porphyrins and porphyrin metal complexes play a fundamental role in many biochemical processes, we believe that it is important to make use of theory to calculate the global and local reactivity parameters for porphyrin and substituted porphyrin system in order to carry out a more detailed analysis of the different effects that have an influence on the chemical reactivity at the carbon atom and at the N atom in terms of a local version of the HSAB principle [14]. The failure [15] of the HSAB principle [14] in case of biological environments of metal ions is already reported. Although there are a good number of literatures [16, 17] on the global parameter study of the porphyrins, a very few [18] are found for the study of the local parameters. But no attempt is found for the study of the local softness and Fukui function on porphyrin. Thus, we feel it necessary to make a comparative study of the quantum chemical local reactivity parameters for better understanding of the preferred sites for coordination with the metal ion (electron acceptor) of the porphyrins. We have invoked semi-empirical AM1 method [19–21] to analyze the charge distribution on the different atomic sites of porphyrin in this study.

## THE GLOBAL REACTIVITY PARAMETERS

Parr, Donnely, Levy, and Palke [22] discovered a new fundamental quantity as a new index of chemical reactivity known as the electronic chemical potential ($\mu$). The chemical potential ($\mu$) is a characteristic property of atoms, molecules, ions, and radicals and is the first derivative of energy with respect to the number of electron. Parr et al. [22] showed that the slope, $[\partial E(r)/\partial N]_v$ of the Energy E(r) versus the number of electrons (N) curve at a constant external potential(v), is the chemical potential, $\mu$, and this property, like thermodynamic chemical potential [23] measures the escaping tendency of electrons in the species. Then following Iczkowski and Margrave [24], Parr et al. [22] defined the electronegativity as the additive inverse of the chemical potential:

$$\chi = -\mu \tag{1}$$

or,
$$\chi = -[\partial E/\partial N]_v \tag{2}$$

Parr and Pearson [25], within the scope of the density functional theory (DFT), have rigorously defined the term hardness as the second order derivative of energy with respect to the number of electron that is,

$$\eta = 1/2 \ \{(\delta^2 E/\delta N^2)_v\} \tag{3}$$

Invoking finite difference approximation, Parr and Pearson [25] gave approximate and operational formulae for electronegativity and hardness as under:

$$\chi = (I + A)/2 \tag{4}$$
$$\eta = (I - A)/2 \tag{5}$$

where I and A are the first ionization potential and electron affinity of the chemical species.

The softness(S) is the reciprocal of the hardness;

$$S = 1/\eta \tag{6}$$

According to Koopmans' theorem the orbital energies of the Frontier Orbitals can be written as:

$$-\varepsilon_{HOMO} = I \tag{7}$$

and

$$-\varepsilon_{LUMO} = A. \tag{8}$$

In 1986, within the limitations of Koopmans' theorem, Pearson [26] putted electronegativity and hardness into a MO framework as follows:

$$\chi = -(\varepsilon_{LUMO} + \varepsilon_{HOMO})/2 \tag{9}$$

and

$$\eta = (\varepsilon_{LUMO} - \varepsilon_{HOMO})/2 \tag{10}$$

He again pointed out that a hard species has a large HOMO-LUMO gap and a soft species has a small HOMO-LUMO gap [26].

The electrophilicity index, $\omega$ is a descriptor of reactivity that allows a quantitative classification of the global electrophilic nature of a molecule within a relative scale. Parr and Yang [27] suggested that electronegativity squared divided by hardness measures the electrophilic power of a ligand its prosperity to "soak up" electrons.

Thus,

$$\omega = \mu^2/2\eta \tag{11}$$

It is further anticipated that electrophilicity index should be related to electron affinity, because both electrophilicity index and electron affinity measures capacity of an agent to accept electrons. Electron affinity reflect capability of an agent to accept only one electron from the environment, whereas electrophilicity index measures the energy lowering of a ligand due to maximal electron flow between the donor and acceptor. The electron flows may be either less or more than one. Thus the electrophilicity index provides the direct relationship between the rates of reaction and the electrophilic power of the inhibitors [28].

## THE LOCAL REACTIVITY PARAMETERS

The Fukui functions play a prominent role in the field known as conceptual Density Functional Theory (DFT) [29]. Yang and Parr [30] based on the original ideas of Fukui, Yonezawa, and Shingu [31], introduced Fukui function which reflect the response of a molecular system towards a change in the number of electrons (N) of the molecular system under consideration. The Fukui functions are a measure of local reactivity and defined as:

$$f(r) = (\partial \rho (r)/\partial N)_v. \tag{12}$$

where $\rho(r)$ is the electron density.

It is transparent from the above equation that the Fukui functions measured the response of the electron density at every point r, in front of a change in the number of electrons, N under the constant external potential, v. The sites with the largest value for the Fukui functions are those with the largest response, and as such the most reactive sites within a molecule. In chemistry, often chemical reactivity and molecular properties in general are preferably interpreted in terms of the atoms composing molecular structure. It is then logical to introduce the so called atom condensed Fukui functions. This means that some way of calculating the change in the total atomic electron density of an atom "a" with respect to N is needed. Since the nuclear change of an atom is a constant, one of the easiest ways is to use the concept of atomic charges, which introduces the following expression for atom condensed Fukui function:

$$f_\alpha = - (\partial q_a/\partial N)v \tag{13}$$

Yang and Mortier [32] were the first to use such atom condensed Fukui functions, and used Mulliken charges to obtain values for the above defined atom condensed Fukui functions.

Now, let us discuss the operational definitions of Fukui functions and local softnesses.

In Frontier Orbital theory, the two orbitals, the HOMO, and the LUMO, are the most important in correlating the molecular reactivity and suggesting orientation of a group in the molecule. It is a fact that the reaction takes between two reactants locally and not globally. The algorithms used to define the Fukui functions are as follows:

for governing electrophilic attack,

$$f^- (r) = [\partial\rho(r)/\partial N]^-_{v(r)} \tag{14}$$

for governing nucleophilic attack,

$$f^+ (r) = [\partial\rho(r)/\partial N]^+_{v(r)} \tag{15}$$

for governing radical attack,

$$f^0 (r) = [\partial\rho(r)/\partial N]^0_{v(r)} \tag{16}$$

Fukui function measures the response of the electron density at every point r and the sites with the largest value for the Fukui functions are those with the largest response, and as such the most reactive sites within the molecule. According to the "frozen core" approximation, the operational algorithms proposed [4] to calculate the Fukui functions are as follows:

for governing electrophilic attack,

$$f^- (r) \approx \rho_{HOMO} (r) \tag{17}$$

for governing nucleophilic attack,

$$f^+ (r) \approx \rho_{LUMO}(r) \tag{18}$$

for governing radical attack,

$$f^0 (r) \approx \tfrac{1}{2} [\rho_{HOMO} (r) + \rho_{LUMO}(r)] \tag{19}$$

The local softness parameters with their specific use are as follows:

The $s^- (r)$ is for governing electrophilic attack,

$$S^- (r) = S\, f^-(r) \tag{20}$$

The $s^+ (r)$ is for governing nucleophilic attack

$$s^+(r) = S\, f^+ (r) \tag{21}$$

The $s^0 (r)$ is for governing radical attack

$$s^0(r) = S\, f^0 (r) \tag{22}$$

The Fukui function has been used in several works as a natural descriptor of site selectivity. Within the Li and Evans reactivity and selectivity rules [33], for soft–soft interactions, the preferred reactive site in a molecule should have the highest value of the Fukui function, whereas the hard–hard interactions are supposed to be described through the minimum value of this local index often the reactivity in molecules with only one reactive site can be correctly characterized. However, for poly-functional systems where more than one site can be attacked, the Fukui function seems to fail predicting the selectivity of hard–hard interactions since hard–hard interactions are charge controlled and soft–soft interactions are Frontier controlled, the Fukui function is not expected to describe well the hard–hard interactions. In a recent work, using the properties of Fukui function, more powerful descriptor of chemical reactivity and site selectivity have been proposed by Chattaraj, Maity, and Sarkar [34].

## THE ATOMIC CHARGE

Mulliken charges [35, 36] provide a means of estimating partial atomic charges from calculations carried out by the methods of computational chemistry, particularly those based on the linear combination of atomic orbitals molecular orbital method. The charge thus arises from the Mulliken population analysis.

Let us consider $C_{\mu i}$ is the coefficients of the basis functions in the molecular orbital for the $\mu$th basis function in the ith molecular orbital. Then the density matrix terms are given by:

$$D_{\mu v} = 2\, C^{*}_{vi} \qquad (23)$$

For a closed shell system where each molecular orbital is doubly occupied the population matrix P has terms:

$$P_{\mu v} = D_{\mu v}\, S_{\mu v} \qquad (24)$$

where S is the overlap matrix of the basis functions and the sum of all terms of $P_{\mu v}$ is N—the total number of electrons.

The Mulliken population analysis aims first to divide N among all the basis functions. This is done by taking the diagonal element of $P_{\mu v}$ and then dividing the off-diagonal elements equally between the two appropriate basis functions. Since the off-diagonal terms include $P_{\mu v}$ and $P_{\mu v}$, this simplifies to just the sum of a row. This defines the gross orbital population (GOP) as

$$(GOP)_{\mu} = \Sigma_{v} P\mu v \qquad (25)$$

The (GOP)$\mu$ terms sum to N and thus divide the total number of electrons between the basis functions. It remains to sum these terms over all basis functions on a given atom A to give the gross atom population (GAP). The sum of $(GAP)_{A}$ terms is also N. The charge, $Q_{A}$, is then defined as the difference between the number of electrons on the isolated free atom, which is the atomic number $Z_{A}$, and the gross atom population:

$$Q_{A} = Z_{A} - (GAP)_{A} \qquad (26)$$

The problem with this approach is the equal division of the off-diagonal terms between the two basis functions. This leads to charge separations in molecules that are exaggerated. In a modified [36] Mulliken population analysis, this problem can be reduced by dividing the overlap populations $P_{\mu v}$ between the corresponding orbital populations $P_{\mu\mu}$ and $P_{vv}$ in the ratio between the latter. This choice, although still arbitrary, relates the partitioning in some way to the electronegativity difference between the corresponding atoms. Numerous approximations were being tried out for getting around the problem of computing the more difficult integrals. In the zero differential overlap (ZDO) approximation in which the product of two different atomic orbitals is set to zero. The integral which survived the ZDO approximation were partly computed using the uniformly charged sphere approximation and the rest parameterized. The result procedure was a quantitative theory, which went well beyond Hückel theory by explicitly taking into account electron repulsions. Pariser and Parr [37] used the method for the prediction of the spectral procedure of conjugate systems. Pople [38] independently used the ZDO approximation to work out the same computational strategy.

Now, let us have a look on the zero differential overlap approximation [21] to pound over the subject.

It is well known in quantum chemistry that for two different atomic functions $\chi_{\mu}$ and $\chi_{v}$ the overlap integral is:

$$S_{\mu v} = \int \chi_{\mu}\, \chi_{v}\, d\tau \;\; (\mu \neq v) \qquad (27)$$

The differential overlap between these two function which is simply the product $\chi_\mu$ (i) $\chi_v$ (ii) gives the probability of finding an electron i in a common volume element to them. It can be expressed as:

$$\chi_\mu (i)\, \chi_v (ii) = \delta_{\mu v} \tag{28}$$

If $\chi_\mu$ and $\chi_v$ are centered on two atoms distant from each other or their spatial orientations are quite different their differential overlap is nearly zero. All approximate LCAO-MO-SCF schemes have made use of this feature by neglecting all or most of the integrals containing the product $\chi_\mu (i)\, \chi_v (ii)$ unless $\mu$ is equal to v. If all such integrals are neglected we come at the so—called zero differential overlap (ZDO) approximation in which $\delta_{\mu v}$ is set equal to zero unless $\mu$ and v are equal. Actually orbitals which are centered on the same or even on two directly bonded atoms have a rather larger differential overlap. Setting such differential overlap equal zero is far from being a correct approximation. As mentioned above the essential advantage of applying the ZDO approximation is the considerable simplification it introduces in the computation. The capability of the ZDO methods to describe the different molecular properties is an important measure of the appropriateness of the ZDO approximation. Practical experience with different ZDO methods has proved that they can be brought about to reasonably represent chemical properties. However it is not expected that ZDO methods may provide calculated values which are in good agreement with rigorous ab initio calculations. There have been several attempts [39–41] to show that utilizing the ZDO approximation is theoretically justified.

## METHOD OF COMPUTATION

We have applied semi-empirical Austin Model 1 (AM1) [19–21] for the study of the electronic structure and chemical reactivity of porphyrin. The AM1 calculations are carried out using ArgusLab4.0 software [42]. The structure of porphyrin was initially taken as planar, and the geometry optimization was done using Hartree Fock SCF method. The optimized structure and the HOMO and LUMO charge densities are used in the Figs. 9.1–9.3 respectively to understand the most stable (less repulsion) configuration and electronic distribution of the molecules. The eigen values are for the highest occupied molecular orbital (HOMO) and the lowest unoccupied molecular orbital (LUMO) of the porphyrin molecule are taken as the HOMO and LUMO orbital energy. Using these orbital energies, we have computed the global reactivity descriptors such as $\mu$, $\chi$, $\eta$, S, and $\omega$ for porphyrin following the equations (1), (4–6), and (11) respectively. We have reported the physical properties of porphyrin and presented them in Table 9.1. We have computed the global reactivity parameters of the molecule and presented them in Table 9.2. The Fukui functions and local softnesses for each site of the porphyrin molecule are computed using the equations (17–19) and equations (20–22) respectively and presented them in Table 9.3. In Table 9.4, we have summarized the computed Mulliken and ZDO charge on different atomic sites of porphyrin.

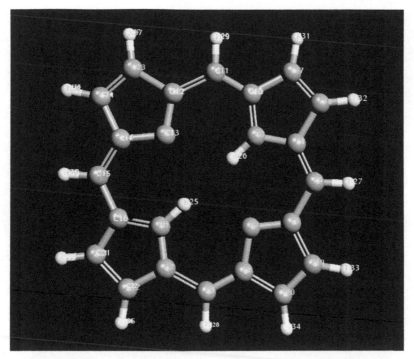

**Figure 9.1.** Optimized structure of porphyrin.

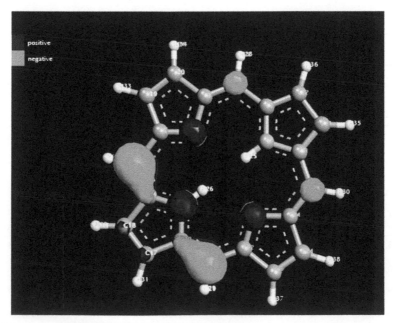

**Figure 9.2.** Charge density on HOMO of porphyrin molecule.

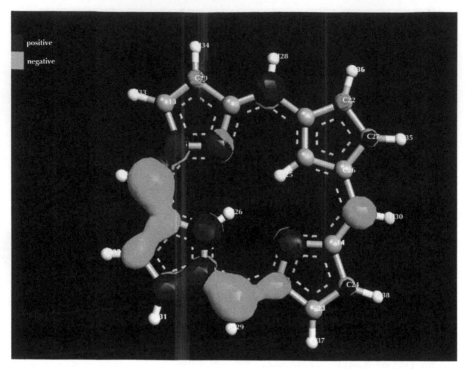

**Figure 9.3.** Charge density on LUMO porphyrin molecule.

**Table 9.1.** Physical properties of porphyrin.

| System | Energy (SCF) (au) | Energy (Geometrical) (au) | Heat of Formation (kcal/mol) | Dipole moment (Debye) |
|---|---|---|---|---|
| Porphyrin | −131.26243954 | −131.3497128 | 245.6856 | 1.07522378 |

**Table 9.2.** The global reactivity parameters.

| System | I(au) | A(au) | μ(au) | χ(au) | η(au) | S(au) | ω(au) |
|---|---|---|---|---|---|---|---|
| Porphyrin | 0.287271 | 0.052436 | −0.16985 | 0.16985 | 0.117475 | 8.51661805 | 0.001695 |

**Table 9.3.** Calculated Fukui functions (au) and local softnesses (au) of porphyrin.

| Atom no. | $f^+(r)$ | $f^-(r)$ | $f^0(r)$ | $s^+$ | $s^-$ | $s^0$ |
|---|---|---|---|---|---|---|
| 1 N | $3.9469 \times 10^{-7}$ | $1.46638 \times 10^{-7}$ | $2.70664 \times 10^{-7}$ | $3.36143978 \times 10^{-6}$ | $1.248859 \times 10^{-6}$ | $2.3051419 \times 10^{-6}$ |
| 2 C | 0.016913 | 0.364502 | 0.1907075 | 0.144041 | 3.104324 | 1.624183 |
| 3 N | $9.5948 \times 10^{-8}$ | $1.9727 \times 10^{-7}$ | $1.46609 \times 10^{-7}$ | $8.17152 \times 10^{-7}$ | $1.68007 \times 10^{-6}$ | $1.2486128 \times 10^{-6}$ |
| 4 C | 0.169308 | 0.052944 | 0.111126 | 1.441932 | 0.450902 | 0.946418 |

**Table 9.3.** *(Continued)*

| Atom no. | f⁺(r) | f⁻(r) | f⁰(r) | s⁺ | s⁻ | s⁰ |
|---|---|---|---|---|---|---|
| 5 C | 0.171002 | 0.002894 | 0.086948 | 1.456359 | 0.024649 | 0.740503 |
| 6 C | 0.07863 | 0.100753 | 0.0896915 | 0.669662 | 0.858073 | 0.763868 |
| 7 N | 0.135359 | 0.004294 | 0.0698265 | 1.152802 | 0.036571 | 0.594686 |
| 8 C | 0.006047 | 0.237245 | 0.121646 | 0.0515 | 2.020525 | 1.036013 |
| 9 C | 0.155013 | 0.041568 | 0.0982905 | 1.320184 | 0.354015 | 0.837103 |
| 10 C | 0.169295 | 0.053095 | 0.111195 | 1.441821 | 0.45219 | 0.947005 |
| 11 C | 0.171007 | 0.002985 | 0.086996 | 1.456399 | 0.02542 | 0.740912 |
| 12 C | 0.078612 | 0.100831 | 0.543461 | 0.66951 | 8.587394 | 4.62845 |
| 13 N | 0.135347 | 0.00436 | 0.0698535 | 1.1527 | 0.037132 | 0.594916 |
| 14 C | 0.006042 | 0.237391 | 0.1217165 | 0.051455 | 2.021773 | 1.036613 |
| 15 C | 0.154979 | 0.041732 | 0.0983555 | 1.319894 | 0.355412 | 0.837656 |
| 16 C | 0.016979 | 0.36446 | 0.1907195 | 0.1446 | 3.103969 | 1.624285 |
| 17 C | 0.173494 | 0.007175 | 0.0903345 | 1.477584 | 0.061106 | 0.769344 |
| 18 C | 0.173483 | 0.007223 | 0.090353 | 1.477491 | 0.061512 | 0.769502 |
| 19 C | 0.006579 | 0.054298 | 0.0304385 | 0.05603 | 0.462438 | 0.259233 |
| 20 C | 0.035638 | 0.004007 | 0.0198035 | 0.303192 | 0.034124 | 0.168659 |
| 21 C | 0.051992 | 0.12998 | 0.090986 | 0.442792 | 1.10699 | 0.774893 |
| 22 C | 0.052005 | 0.128832 | 0.0904185 | 0.442909 | 1.097211 | 0.77006 |
| 23 C | 0.006575 | 0.054517 | 0.030546 | 0.055995 | 0.464297 | 0.260149 |
| 24 C | 0.035628 | 0.004008 | 0.019818 | 0.30343 | 0.034132 | 0.168782 |

**Table 9.4.** Computed Mulliken and ZDO charge of porphyrin.

| Atom | Mulliken Charges | Atomic ZDO Atomic charge | Atom | Mulliken Atomic Charges | ZDO Atomic charge |
|---|---|---|---|---|---|
| 1 N | −0.2449 | −0.0967 | 20 C | −0.2005 | −0.1353 |
| 2 C | −0.043 | −0.058 | 21 C | −0.2216 | −0.1662 |
| 3 N | −0.3918 | −0.2507 | 22 C | −0.2216 | −0.1662 |
| 4 C | 0.1044 | 0.0905 | 23 C | −0.2472 | −0.1829 |
| 5 C | −0.2314 | −0.1878 | 24 C | −0.2005 | −0.1353 |
| 6 C | 0.0474 | 0.0356 | 25 H | 0.3744 | 0.2935 |
| 7 N | −0.2389 | −0.2031 | 26 H | 0.3751 | 0.2938 |
| 8 C | −0.0818 | −0.0865 | 27 H | 0.2113 | 0.1469 |

Table 9.4. (Continued)

| Atom | Mulliken Charges | Atomic ZDO Atomic charge | Atom | Mulliken Atomic Charges | ZDO Atomic charge |
|------|------|------|------|------|------|
| 9 C | −0.0499 | −0.0019 | 28 H | 0.2077 | 0.1436 |
| 10 C | 0.1044 | 0.0905 | 29 H | 0.2112 | 0.1469 |
| 11 C | −0.2314 | −0.1878 | 30 H | 0.2077 | 0.1436 |
| 12 C | 0.0474 | 0.0356 | 31 H | 0.2304 | 0.1615 |
| 13 N | −0.2389 | −0.2031 | 32 H | 0.2304 | 0.1615 |
| 14 C | −0.0818 | −0.0865 | 33 H | 0.2192 | 0.1527 |
| 15 C | −0.0499 | −0.0019 | 34 H | 0.2175 | 0.1515 |
| 16 C | −0.043 | −0.058 | 35 H | 0.2197 | 0.1534 |
| 17 C | −0.1994 | −0.1339 | 36 H | 0.2197 | 0.1534 |
| 18 C | −0.1994 | −0.1339 | 37 H | 0.2192 | 0.1527 |
| 19 C | −0.2472 | −0.1829 | 38 H | 0.2175 | 0.1515 |

A look on the Table 9.3 reveals that the centers N-1, N-3, and N-7, N-13, and C-8, C-14, and C-9, C-15, and C-5, C-11,C-17, C-18, and C-4, C-10, and C-6, C-12, and C-2, C-16, and C-19, C-23, and C-20, C-24, and C-21, C-22 are almost equal in their $f^+(r)$, $f^0(r)$, and $f^-(r)$ value. The local softnesses of the centers are also following the similar trend of variation. A deeper study on the Table 9.3 reveals that the center N-1, N-3 are lowest and C-5, C-11, C-17, C-18 have the highest $f^+(r)$ and $s^+$ value and the trans pair, N-1 and N-3 have the lowest and C-2, C-16 have the highest $f^-(r)$ and $s^-$ value.

Thus, it is clear from the above discussions that the local softness and Fukui functions are not useful for the study of the bio-interaction between the donor sites of the porphyrin ring and the metal ion. The failure of the Fukui function for the study of the softest and hardest regions of some soft and hard molecule was reported by Torrent-Sucarrat, Proft, Geerlings, and Ayers [43].

It is well known to us that the four coordination number of Fe (II) (in case of hemoglobin) is satisfied by four N atoms of the macromolecule, which are not the maximum reactive site according to their Fukui function and local softness values in that molecule. We found the suggestion of Mendez and Gazquez [14] where it was stated that the interaction between two molecules will occur not necessarily through their softest atoms but rather through those atoms whose Fukui functions are the same. Parr and Yang [27] previously showed that the interactions between two molecules say A and B having atoms, $a_1$, $a_2$ and so on and $b_1$, $b_2$, and so on does not necessarily occur through the softest atom of A and B (say $a_1$ for A and $b_1$ for B) but through those atom of A and B (say $a_2$ for A, and $b_2$ for B) whose Fukui functions [say $f(a_2)$ and $f(b_2)$] are approximately close to each other, that is, $f(a_2) \approx f(b_2)$. If we consider the example of heme where the metal Fe(II) is a borderline acid having some intermediate hardness value between hard acids and soft acids, we can pointed out that our computed results (Table 9.3) show that the N-atoms have some intermediate $f^+$ value. This local version of HSAB principal was applied to rationalize the regioselectivity in several Diels–Alder reaction [44]. Mendez and Gazquez [14] pointed out that this interaction can be proved by the minimization of the grand canonical potential without assuming the local softness equalization that is, $S_A = S_B$. This is a general statement of local HSAB principal to determine the atom through which the interaction between A and B takes place. The regions of a molecule where the Fukui function is large are chemically softer than the region where the Fukui function is small, and by invoking the HSAB principal in a local sense, one may establish the behavior of the different sites with respect to hard or soft reagents. But in this type of work, where one concentrate only to the ligand and try to examine the electronic distribution on the atoms in the molecule, local HSAB principle [14] cannot be applied because the Fukui functions for the acceptor cannot be computed in the same computational environment.

Although there are views [43] that the Local HSAB principle [14] cannot be applied to study the biological systems and their functions, we can point out that the interaction of the border line acids $Fe^{2+}$ will takes place through the borderline local centers (the N-atoms) of the porphyrin ring, but this is not enough explanation because this conclusion is too qualitative.

Thus we can say that both the Fukui functions and local softnesses fail to explain the maximum reactive or more precisely the donor site of the molecule/ligand porphyrin. But when we look on the pi charge density, ZDO and Mulliken charges on the atomic sites we surprisingly noted that two sets of the trans N atoms of the molecule has the same value and higher in magnitude than all C atoms. Again one set of trans N atoms has the highest value. We know that in porphyrin, two trans sets are initially different, one set contains two non-protonated N atom and other contains two protonated N atoms (N-H). This is due to the fact that in case of the non-protonated free lone pair of electron is present while in the case of other set, the lone pair is utilized for $H^+$.

## CONCLUSION

In this work, on the basis of semi-empirical AM1 model, we have calculated the eigen values and eigen functions of molecules. We have also calculated the Global DFT descriptors for porphyrins. The Fukui functions, local softness, and atomic charges for each center of the porphyrin are also calculated. We have found that all the centers are not equal in chemical reactivity. The differences in chemical reactivity of different sites for are nicely represented in this work. It is well known fact that the host porphyrins bind the guest into their cavity. But this fact cannot be predicted by the Fukui function and the local softness values. From the above discussions it is distinct that the largest values of the Fukui function and local softness do not necessarily correspond to the softest regions of the molecule. Based on our results, we can conclude that it is more useful to interpret the ZDO atomic charge and Mulliken charge as function that measure the local abundance or concentration of the charges on the atom in the molecule. The purpose of mathematical definitions of the two very important conceptual density functional reactivity descriptor, the Fukui function, and local softness is to provide a quantitative representation of qualitative concepts. One goal of this work is to show that the explanation that the largest value of the Fukui function and local softness allow one to identify the softest and hardest center of a molecule is always not correct always. Thus further work is certainly warranted. In particular, the conditions under which mathematical and the chemical formulations of the two parameters coincide need to be explored. Most ambitiously the density functional theorist should examine whether there may be alternative mathematical definition for them that more faithfully model the eponymous chemical concepts.

## KEYWORDS

- **Fukui functions**
- **Ligand**
- **Macrocycle**
- **Metallocomplexes**
- **Porphyrins**
- **Tetraphenylporphyrin**

# References

## 1

1. Jensen, W.B. (1996). Electronegativity from Avogadro to Pauling. *J. Chem. Educ.*, **73**, 11–20.

2. Avogadro, A. (1809). Idees sur Iacidite et Ialcalinite. *J. Phys. Chim. Hist. Nat.*, **69**, 142–148.

3. Morselli, M. (1984). *Amedeo Avogadro: A scientific biography*. Kluwer, Dordrecht.

4. Pierson, S. (1984). Avogadro and his work: Amedeo Avogadro, *Science*, **226**, 432–433.

5. Berzelius, J.J. (1813). An explanatory statement of the notions or principles upon which the systematic arrangement is founded which was adopted as the basis of an essay on chemical nomenclature. *J. Nat. Phil. Chem. Arts* (Nicolson's J.), **34**, 142–146, 240–246, 313–319.

6. Berzelius, J.J. (1819). *Essai Sur La Theorie Des Proportions Chimiques Et Sur Linfluence Chemique* (French Edition). Kessinger Publishing, Paris.

7. Jorpes, J.E. (1966). *Jac Berzelius: His life and work*. Almqvist and Wiklsell, Stockholm.

8. Melhado, E.M. (1981). *Jacob Berzelius: The emergence of his chemical system*. University of Wisconsin, Madison,WI.

9. Russell, C.A. (1963). The electrochemical theory of Berzelius. *Ann. Sci.*, **19**, 117–145.

10. Pauling, L. (1932). The nature of the chemical bond IV. The energy of single bonds and the relative electronegativity of atoms. *J. Am. Chem. Soc.*, **54**, 3570–3582.

11. Pauling, L. (1960). *The nature of chemical bond*. Cornell University Press, Ithaca.

12. Allred, A.L. and Rochow, E.G. (1958). Electronegativity of carbon, silicon, germanium, tin and lead. *J. Inorg. Nucl. Chem.*, **5**, 269–288.

13. Allen, L.C. (1989). Electronegativity is the average one-electron energy of the valence-shell electrons in ground-state free atoms. *J. Am. Chem. Soc.*, **111**, 9003–9014.

14. Allen, L.C. (1992). Extension and completion of the periodic table. *J. Am. Chem. Soc.*, **114**, 1510–1511.

15. Huheey, J.E., Keiter, E.A., and Keiter, R.L. (1983). *Inorganic chemistry: Principles of structure and reactivity*, 3rd ed. Harper & Row, New York.

16. Frenking, G. and Krapp, A. (2007). Unicorns in the world of chemical bonding models. *J. Comput. Chem.*, **28**, 15–24.

17. Ghosh, D.C. and Islam, N. (2010). A quest for the algorithm for evaluating the molecular hardness. *Int. J. Quantum Chem.*, DOI: 10.1002/qua.22499 [Early View].

18. Ghosh, D.C. and Islam, N. (2010). Whether there is a hardness equalization principle analogous to the electronegativity equalization principle—A quest. *Int. J. Quantum Chem.*, DOI: 10.1002/qua.22508 [Early View].

19. Ghosh, D.C. and Islam, N. (2010). Determination of some descriptors of the real world working on the fundamental identity of the basic concept and the origin of the electronegativity and the global hardness of atoms. Part-1, Evaluation of internuclear bond distance of some heteronuclear diatomics, *Int. J. Quantum Chem.*, DOI: 10.1002/qua.22500 [Early View].

20. Ghosh, D.C. and Islam, N. (2010). Determination of some descriptors of the real world working on the fundamental identity of the basic concept and the origin of the electronegativity and the global hardness of atoms. Part-2. Computation of the dipole moments of some heteronuclear diatomics. *Int. J. Quantum Chem.*, DOI: 10.1002/qua.22651 [Early View].

21. Ghosh, D.C. and Islam, N. (2010). Charge transfer associated with the physical process of hardness equalization and the chemical event of the molecule formation and the dipole moments. *Int. J. Quantum Chem.*, DOI: 10.1002/qua.22653 [Early View].

22. Ghosh, D.C. and Islam, N. (2011). Whether electronegativity and hardness are manifest

two different descriptors of the one and the same fundamental property of atoms: A quest. *Int. J. Quantum Chem.*, **111**, 40–51.

23. Fukui, K. (1982). Role of frontier orbitals in chemical reactions. *Science*, **218**, 747–754.

24. Pritchard, H.O. and Skinner, H.A. (1955). The concept of electronegativity. *Chem. Rev.*, **55**, 745–786.

25. Ghosh, D.C. (2003). The scales and concept of electronegativity. *J. Indian Chem. Soc.*, **80**, 527–533.

26. Cherkasov, A.R., Galkin, V.I., Zueva, E.M., and Cherkasov, R.A. (1998). The concept of electronegativity. The current state of the problem. *Russ. Chem. Rev.*, **67**, 375–392.

27. Sen, K.D. and Jorgensen, C.K. (1987). *Electronegativity*. Springer-Verlag, New York.

28. Reddy, R.R., Rao, T.V.R., and Viswanath, R. (1989). Correlation between electronegativity differences and bond energies. *J. Am. Chem. Soc.*, **111**, 2914–2915.

29. Myers, R.T. (1979). Physical and chemical properties and bonding of metallic elements. *J. Chem. Educ.*, **56**, 711–718.

30. Gordy, W. (1951). Interpretation of nuclear quadrupole couplings in molecules, *J. Chem. Phys.*, **19**, 792–793.

31. Gutowsky, H.S. and Hoffman, C.J. (1951). Nuclear magnetic shielding in fluorine and hydrogen compounds. *J. Chem. Phys.*, **19**, 1259–1267.

32. Lackner, K.S. and Zweig, G. (1983). Introduction to the chemistry of fractionally charged atoms: Electronegativity. *Phys. Rev. D.*, **28**, 1671–1691.

33. Wahl, U., Rita, E., Correia, J.G., Marques, E., Alves, A.C., and Soares, J.C. (2005). Direct evidence for As as a Zn-site impurity in ZnO. *Phys. Rev. Lett.*, **95**, 215503–215506.

34. Asokamani, R. and Manjula, R. (1989). Correlation between electronegativity and superconductivity. *Phys. Rev. B.*, **39**, 4217–4221.

35. Hur, S.G., Kim, T.W., Hwang, S.J., Park, H., Choi, W., Kim, S.J., and Choy, J.H. (2005). Synthesis of new visible light active photocatalysts of Ba (In1/3Pb1/3M1/3')O3 (M' = Nb, Ta): A band gap engineering strategy based on electronegativity of a metal component. *J. Phys. Chem. B.*, **109**, 15001–15007.

36. Zhang, L., Wang, E.G., Xue, Q.K., Zhang, S.B., and Zhang Z.(2006). Generalized electron counting in determination of metal-induced reconstruction of compound semiconductor surfaces. *Phys. Rev. Lett.*, **97**, 126103–126106.

37. Zueva, E.M., Galkin, V.I., Cherkasov, A.R., and Cherkasov, R.A. (2002). Electronic chemical potential and orbital electronegativity of univalent substituents. *Russ. J. Org. Chem.*, **38**, 624–631.

38. Arroyo-De Dompablo, M.E., Armand, M., Tarascon, J.M., and Amador, U. (2006). On-demand design of polyoxianionic cathode materials based on electronegativity correlations: An exploration of the Li$_2$MSiO$_4$ system (M = Fe, Mn, Co, Ni). *Electrochem. Commun.*, **8**, 1292–1298.

39. Kobayashi, S., Hamashima, H., Kurihara-Miyata, N., and Tanaka, A. (1998). Hardness controlled enzymes and electronegativity controlled enzymes: Role of an absolute hardness-electronegativity (η-X) activity diagram as a coordinate for biological activities. *Chem. Pharm. Bull.*, **46**, 1108–1115.

40. Islam, N. and Ghosh, D.C. (2010). Evaluation of global hardness of atoms based on the commonality in the basic philosophy of the origin and the operational significance of the electronegativity and the hardness. Part-1. The gordy's scale of electronegativity and the global hardness. *Eur. J. Chem.*, **1**, 83–89.

41. Islam, N. and Ghosh, D.C. (2011). A new radial dependent electrostatic algorithm for the evaluation of the electrophilicity indices of the atoms. *Int. J. Quantum Chem.*, DOI: 10.1002/qua.22861 [Early View].

42. Haissinsky, M. (1946). Échelle des électronégativités de Pauling et chaleurs de formation des composés inorganiques. *J. Phys. Radium.*, **7**, 7–11.

43. Huggins, M.L. (1953). Bond energies and polarities. *J. Am. Chem. Soc.*, **75**, 4123–4126.

44. Gordy, W. and Orville Thomas, W.J. (1956). Electronegativities of the elements. *J. Chem. Phys.*, **24**, 439–444.

45. Altshuller, A.P. (1954). The electronegativities and some electron affinities of copper, zinc, and gallium subgroup elements. *J. Chem. Phys.*, **22**,765–765.

46. Allred, A.L. (1961). Electronegativity values from thermochemical data. *J. Inorg. Nucl. Chem.*, **17**, 215.

47. Mulliken, R.S. (1935) Electronic structures of molecules XI. Electroaffinity, molecular orbitals and dipole moments. *J. Chem. Phys.*, **3**, 573–585.

48. Putz, M.V. (2006). Systematic formulations for electronegativity and hardness and their atomic scales within density functional softness theory. *Int. J. Quantum Chem.*, **106**, 361–389.

49. Putz, M.V. (2008). *Absolute and chemical electronegativity and hardness*. Nova Science Publishers, New York.

50. Ghosh, D.C. (2005). A new scale of electronegativity based on absolute radii of atoms. *J. Theoret. Comput. Chem.*, **4**, 21–33.

51. Ghosh, D.C. and Gupta, K. (2006). A new scale of electronegativity of 54 elements of periodic table based on polarizability of atoms. *J. Theoret. Comput. Chem.*, **5**, 895–911.

53. Ghosh, D.C. and Chakraborty, T. (2009). Gordy's electrostatic scale of electronegativity revisited. *J. Mol. Struct.* (THEOCHEM), **906**, 87–93.

52. Ghosh, D.C., Chakraborty, T., and Mandal, B. (2009). Allred Rochow's scale of electronegativity revisited. *Theor. Chem. Account.*, **124**, 295–301.

54. Malone, J.G. (1933). The electric moment as a measure of the ionic nature of covalent bonds. *J. Chem. Phys.*, **1**, 197–199.

55. Mulliken, R.S. (1934). A new electroaffinity scale: Together with data on valence states and on valence ionization potentials and electron affinities. *J. Chem. Phys.*, **2**, 782–793.

56. Hund, F. (1931). Zur Frage der chemischen Bindung. *Zeits. f. Physik*, **73**, 1–30.

57. Coulson, C.A. (1951). Critical survey of the method of ionic-homopolar resonance. *Proc. Roy. Soc. London A*, **207**, 63–73.

58. Parr, R. G., Donnelly, R.A., Levy, M., and Palke, W.E. (1978). Electronegativity, the density functional viewpoint. *J. Chem. Phys.*, **68**, 3801–3807.

59. Parr, R.G. and Pearson, R.G. (1983). Absolute hardness: Companion parameter to absolute electronegativity. *J. Am. Chem. Soc.*, **105**, 7512–7516.

60. Bratsch, S.G. (1988). Revised Mulliken electronegativities: I. Calculation and conversion to Pauling units. *J. Chem. Educ.*, **65**(1), 34–41.

61. Bratsch, S.G. (1988) Revised Mulliken electronegativities: II. Applications and limitations. *J. Chem. Educ.*, **65**(3), 223–227.

62. Hinze, J. and Jaffe, H.H. (1962). Electronegativity. I. Orbital Electronegativity of neutral atoms. *J. Am. Chem. Soc.*, **84**, 540–546.

63. Hinze, J., Whitehead, M.A., and Jaffe, H.H. (1963). Electronegativity. II. Bond and orbital electronegativities. *J. Am. Chem. Soc.*, **85**(2), 148–154.

64. Pearson, R.G. (1963). Hard and soft acids and bases. *J. Am. Chem. Soc.*, **85**, 3533–3539.

65. Gordy, W. (1946). A new method of determining electronegativity from other atomic properties. *Phys. Rev.*, **69**, 604–607.

66. Pasternak, A. (1977). Electronegativity based on the simple bond charge model. *Chem. Phys.*, **26**, 101–112.

67. Ray, N.K., Samuels, L., and Parr, R.G. (1979). Study of electronegativity equalization. *J. Chem. Phys.*, **70**, 3680–3684.

68. Politzer, P., Parr, R.G., and Murphy, D.R. (1983). Relationships between atomic chemical potentials, electrostatic potentials, and covalent radii. *J. Chem. Phys.*, **79**, 3859–3871.

69. Islam, N. (2010). Correlation between the electronegativity ansatz of Mulliken and Gordy. *J. Mole. Struct.* (THEOCHEM), **947**, 123–123.

70. Walsh, A.D. (1951). Factors affecting bond strengths. I. A possible new definition of electronegativity. *Proc. Roy. Soc. London A*, **207**, 13–22.

71. Sanderson, R.T. (1951). An interpretation of bond lengths and a classification of bonds. *Science*, **114**, 670–672.

72. Parr, R.G. and Bartolotti, L.J. (1982). On the geometric mean principle for electronegativity equalization. *J. Am. Chem. Soc.*, **104**, 3801–3803.

73. Parr, R.G. and Yang, W. (1989). *Density functional theory of atoms and molecules.* Oxford University Press, New York.

74. Hohenberg, P. and Kohn, W. (1964) Inhomogeneous electron gas. *Phys. Rev. B.* **136**, 864–871.

75. Gyftopoulos, E.P. and Hatsopoulos, G.N. (1968). Quantum thermodynamic definition of electronegativity. *Proc. Natl. Acad. Sci. USA*, **60**, 786–793.

76. Little, E.J. and Jones, M.M. (1960). A complete table of electronegativities. *J. Chem. Educ.*, **37**, 231–232.

77. Mande, C., Deshmukh, P., and Deshmukh, P. (1977). A new scale of electronegativity on the basis of calculations of effective nuclear charges from X-ray spectroscopic data. *J. Phys. B: At. Mol. Phys.*, 10, 2293–2301.

78. Boyd, R.J. and Markus, G.E. (1981). Electronegativities of the elements from a nonempirical electrostatic model. *J. Chem Phys.*, **75**, 5385–5388.

79. Boyd, R.J. (1977). The relative sizes of atoms. *J. Phys. B*, **10**, 2283–1189.

80. Clementi, E. and Roetti, C. (1974). Roothaan-Hartree-Fock atomic wavefunctions: Basis functions and their coefficients for ground and certain excited states of neutral and ionized atoms, Z ≤ 54. *At. Data Nucl. Data Tables*, **14**, 177–478.

81. Iczkowski, R.P. and Margrave, J.L. (1961). Electronegativity. *J. Am. Chem. Soc.*, **83**, 3547–3551.

82. Klopman, G. (1964). A Semiempirical treatment of molecular Structures. II. Molecular terms and application to diatomic molecules. *J. Am. Chem. Soc.*, **86**, 4550–4557.

83. Yuan, H.J. (1964). Studies on electronegativity- I. The electronegativity of atoms. *Acta Chimica Sinica*, **30**, 341–352.

84. Luo, Y.R. and Benson, S.W. (1988). New electronegativity scale for the correlation of heats of formation. 1. Alkyl derivatives. *J. Phys. Chem.*, **92**, 5255–5257.

85. Luo, Y.R. and Benson, S.W. (1988). Relationships between atomic chemical potentials, electrostatic potentials, and covalent radii. *J. Am. Chem. Soc.*, **111**, 5255–5257.

86. Luo, Y.R. and Pacey, P.D. (1991). Theoretical support for a new electronegativity scale. *J. Am. Chem. Soc.*, **113**, 1465–1466.

87. St. John, A. and Bloch, N. (1974). Quantum-defect electronegativity scale for nontransition elements. *Phys. Rev. Lett.*, **33**, 1095–1098.

88. Bloch, A.N. and Simons, G. (1972). Structural index for elemental solids. *J. Am. Chem. Soc.*, **94**, 8611–8613.

89. Parr, R.G. and Yang, W. (1984). Density functional approach to the frontier-electron theory of chemical reactivity. *J. Am. Chem. Soc.*, **106**, 4049–4050.

90. Pearson, R.G. (1986) Absolute electronegativity and hardness correlated with molecular orbital theory. *Proc. Natl. Acad. Sci.*, **83**, 8440–8441.

91. Zhang, Y.H. (1982). Electronegativities of elements in valence states and their applications. 1. Electronegativities of elements in valence states. *Inorg. Chem.*, **21**, 3886–3889.

92. Zhang, Y.H. (1982). Electronegativities of elements in valence states and their applications. 2. A scale for strengths of Lewis acids. *Inorg. Chem.*, **21**, 3889–3893.

93. Slater, J.C. (1960). *Quantum theory of atomic structure.* Vol. 1, McGraw-Hill, New York.

94. Boyd, R.J. and Edgecombe, K.E. (1988). Atomic and group electronegativities from the electron-density distributions of molecules. *J. Am. Chem. Soc.*, **110**, 4182–4186.

95. Nagle, J.K. (1990). Atomic polarizability and electronegativity. *J. Am. Chem. Soc.*, **112**, 4741–4747.

96. Li, K. and Xue, D. (2006). Estimation of electronegativity values of elements in different valence states. *J. Phys. Chem. A*, **110**, 11332–11337.

97. Noorizadeh, S. and Shakerzadeh, E. (2008). A new scale of electronegativity based on electrophilicity. *J. Phys. Chem. A*, **112**, 3486–3491.

98. Parr, R.G., Szentpaly, L.V., and Liu. S. (1999). Electrophilicity Index. *J. Am. Chem. Soc.*, **121**, 1922–1924.

99. Murphy, L.R., Meek, T.L., Allred, A.L., and Allen, L.C. (2000). Evaluation and test of Pauling's electronegativity scale. *J. Phys. Chem. A*, **104**, 5867–5871.

100. Chattaraj, P.K. and Maity, B. (2001). Electronic structure principles and atomic shell structure. *J. Chem. Educ.*, **78**, 811–813.

101. Chattaraj, P.K., Roy, D.R., and Giri, S. (2007). Electronic structure principles in static and dynamic situations. *Computing Letters* (CoLe), **3**, 223–230.

102. Pearson, R.G. (1968). Failure of Pauling's bond energy equation. *Chem. Commun.* (London), **24**, 65–67.

103. Ayers, P.W. (2007). The physical basis of the hard/soft acid/base principle. *Faraday Discuss*, **135**, 161–190.

104. March, N.H. and White, R.J. (1972). Nonrelativistic theory of atomic and ionic binding energies for large atomic number. *J. Phys. B*, **5**, 466–471.

105. Allen, L.C. and Knight, E.T. (1992). Electronegativity: Why has it been so difficult to define? *J. Mole. Struct: Theochem.*, **261**, 313–330.

106. Allen, L.C. and Huheey, J.E. (1980). The definition of electronegativity and the chemistry of the noble gases. *J. Inorg. Nucl. Chem.*, **42**, 1523–1524.

107. Mann, J.B., Meek, T.L., Knight, E.T., Capitani, J.F., and Allen, L.C. (2000). Configuration energies of the d-block elements. *J. Am. Chem. Soc.*, **122**, 5132–5137.

108. Sanderson, R.T. (1988). Principles of electronegativity Part I. General nature. *Chem. Educ.*, **65**(2), 112–118.

109. Sanderson, R.T. (1988). Principles of electronegativity Part II. Applications. *J. Chem. Educ.*, **65**(3), 227–231.

110. Wu, Z.N. and Sheng, L.G. (1994). Electronegativity: Average nuclear potential of the valence electron. *J. Phys. Chem.*, **98**, 3964–3966.

**2**

1. Pauling, L. (1932). The nature of the chemical bond IV. The energy of single bonds and the relative electronegativity of atoms. *J. Am. Chem. Soc.*, **54**, 3570–3582.

2. Pauling, L. (1960). *The nature of the chemical bond*, 3rd ed., Cornell University Press, Ithaca, NY.

3. Allen, L.C. (1992). Extension and completion of the periodic table. *J. Am. Chem. Soc.*, **114**, 1510–1511.

4. Islam, N. and Ghosh, C.C. (2011). *The time evolution of the electronegativity. Part-1: Concepts and scales, modern trends in chemistry and chemical engineering*, Dr. A.K. Haghi (Ed.), Apple Academic Press, Canada, [submitted].

5. Pritchard, H.O. and Skinner, H.A. (1955). The concept of electronegativity. *Chem. Rev.*, **55**, 745–786.

6. Ghosh, D.C. (2003). The scales and concept of electronegativity. *J. Indian Chem. Soc.*, **80**, 527–533.

7. Cherkasov, A.R., Galkin, V.I., Zueva, E.M., and Cherkasov, R.A., (1998). The concept of electronegativity. The current state of the problem. *Russ. Chem. Rev.*, **67**, 375–392.

8. Coulson, C.A. (1951). Critical survey of the method of ionic-homopolar resonance. *Proc. Roy. Soc. London A*, **207**, 63–73.

9. Sen, K.D. and Jorgensen, C.K. (1987). *Electronegativity*. Springer Verlag, NY.

10. Pearson, R.G. (1989). Absolute electronegativity and hardness: Applications to organic chemistry. *J. Org. Chem.*, **54**, 1423–1430.

11. Sanderson, R.T. (1951). An interpretation of bond lengths and a classification of bonds. *Science*, **114**, 670–672.

12. Parr, R.G., Donnelly, R.A., Levy, M., and Palke, W.E. (1978). Electronegativity, the density functional viewpoint. *J. Chem. Phys.*, **68**, 3801–3807.

13. Parr, R.G. and Bartolotti, L.J. (1982). On the geometric mean principle for electronegativity equalization. *J. Am. Chem. Soc.*, **104**, 3801–3803.

14. Parr, R.G. and Yang, W. (1989). *Density functional theory of atoms and molecules*. Oxford University Press, New York.

15. Hohenberg, P. and Kohn, W. (1964) Inhomogeneous electron gas. *Phys. Rev. B*, **136**, 864–871.

16. Gyftopoulos, E.P. and Hatsopoulos, G.N. (1968). Quantum thermodynamic definition of electronegativity. *Proc. Natl. Acad. Sci. USA*, **60**, 786–793.

17. Ray, N.K., Samuels, L., and Parr, R.G. (1979). Study of electronegativity equalization. *J. Chem. Phys.*, **70**, 3680–3684.

18. Parr, R.G. and Borkman, R.F. (1967). Chemical binding and potential-energy functions for molecules. *J. Chem. Phys.*, **46**, 3683–3685.

19. Parr, R.G. and Borkman, R.F. (1968). Simple bond-charge model for potential-energy curves of homonuclear diatomic molecules. *J. Chem. Phys.*, **49**, 1055–1058.

20. Borkman, R.F. and Parr, R.G. (1968). Toward an understanding of potential-energy functions for diatomic molecules. *J. Chem. Phys.*, **48**, 1116–1126.

21. Pasternak, A. (1977). Electronegativity based on the simple bond charge model. *J. Chem. Phys.*, **26**, 101–112.

22. Huheey, J.E. (1965). The electronegativity of groups. *J. Phys. Chem.*, **69**, 3284–3291.

23. Iczkowski, R.P. (1964). Partial ionic character of diatomic molecules. *J. Am. Chem. Soc.*, **86**, 2329–2332.

24. Pritchard, H.P. (1963). Equalization of electronegativity. *J. Am. Chem. Soc.*, **85**, 1876–1876.

25. Sekhon, B.S. (1998). Equalized electronegativity: Some applications. *Proc. Indian Natl. Sci. Acad. A*, **64**, 581–585.

26. Coulson, C.A. (1953). *Dictionary of values of molecular constants*, Coulson, C.A. and. Daudel, R. (Eds.), Vol. 1. Centre de chimie theorique de France, Paris.

27. Coulson, C.A. and Longuet-Higgnhs, C. (1947). The electronic structure of conjugated systems. I. General theory. *Proc. Roy. Soc. London A*, **191**, 39–60.

28. Mulliken, R.S. (1948). Molecular orbital method and molecular ionization potentials. *Phys. Rev.*, **74**, 736–738.

29. Mulliken, R.S. (1935). Electronic structures of molecules XI. Electroaffinity, molecular orbitals and dipole moments. *J. Chem. Phys.*, **3**, 573–585.

30. Dailey, B.P. and Townes, C.H. (1955). The ionic character of diatomic molecules. *J. Chem. Phys.*, **23**, 118–123.

31. Ghosh, D.C. and Bhattacharya, S. (2005). Computation of quantum mechanical hybridization and dipole correlation of the electronic structure of the $F_3B\text{-}NH_3$ supermolecule. *Int. J. Quant. Chem.*, **105**, 270–279.

32. Pople, J.A., Santry, D.P., and Segal, G.A. (1965). Approximate self-consistent molecular orbital theory. I. Invariant procedures. *J. Chem. Phys.*, **43**, S129–S135.

33. Pople, J.A., Santry, D.P., and Segal, G.A. (1965). Approximate self-consistent molecular orbital theory. II. Calculations with complete neglect of differential overlap. *J. Chem. Phys.*, **43**, S136–S149.

34. Ghosh, D.C. and Chakraborty, A. (2006). Dipole correlation of the electronic structures of the conformations of water molecule evolving through the normal modes of vibrations between angular ($C_2V$) to linear ($D_2h$) shapes. *Int. J. Mol. Sci.*, **7**, 71–96.

35. Nethercot, A.H. Jr. (1978). Molecular dipole moments and electronegativity. *Chem. Phys. Lett.*, **59**, 346–350.

36. Nethercot, A.H. Jr. (1981). Electronegativity and a model hamiltonian for chemical applications. *Chem. Phys.*, **59**, 297–313.

37. Barbe, J. (1983). Convenient relations for the estimation of bond ionicity in A-B type compounds, *J. Chem. Educ.*, **60**, 640–642.

38. Malone, J.G. (1933). The electric moment as a measure of the ionic nature of covalent bonds. *J. Chem. Phys.*, **1**, 197–199.

39. Ghosh, D.C. and Islam, N. (2010). A quest for the algorithm for evaluating the molecular hardness. *Int. J. Quantum Chem.*, DOI: 10.1002/qua.22499, [Early View].

40. Ghosh, D.C. and Islam, N. (2010). Whether there is a hardness equalization principle analogous to the electronegativity equalization principle: A quest. *Int. J. Quantum Chem.*, DOI: 10.1002/qua.22508, [Early View].

41. Ghosh, D.C. and Islam, N. (2010). Determination of some descriptors of the real world working on the fundamental identity of the basic concept and the origin of the electronegativity and the global hardness of atoms.

Part-1, Evaluation of internuclear bond distance of some heteronuclear diatomics. *Int. J. Quantum Chem.*, DOI: 10.1002/qua.22500, [Early View].

42. Ghosh, D.C. and Islam, N. (2010). Determination of some descriptors of the real world working on the fundamental identity of the basic concept and the origin of the electronegativity and the global hardness of atoms. Part-2. Computation of the dipole moments of some heteronuclear diatomics. *Int. J. Quantum Chem.*, DOI: 10.1002/qua.22651, [Early View].

43. Ghosh, D.C. and Islam, N. (2010). Charge transfer associated with the physical process of hardness equalization and the chemical event of the molecule formation and the dipole moments. *Int. J. Quantum Chem.*, DOI: 10.1002/qua.22653, [Early View].

44. Kim, K. (1987). Electronegativity equalization and atomic polar tensor in diatomic molecule. *Bull. Korean Chem. Soc.*, **8**, 432–434.

45. Badger, R.M. (1934). A relation between internuclear distances and bond force constants. *J. Chem. Phys.*, **2**, 128–131.

46. Remick, A.E. (1943). Electronic interpretation of organic chemistry, John Wiley & Sons, Inc., NY.

47. Gordy, W.A. (1946). Relation between bond force constants, bond orders, bond lengths, and the electronegativities of the bonded atoms. *J. Chem. Phys.*, **14**, 305–320.

48. Bratsch, S.G. (1988). Revised Mulliken electronegativities: I. Calculation and conversion to Pauling units. *J. Chem. Edu.*, **65**, 34–41.

49. Fineman, M.A. (1958). Correlation of bond dissociation energies of polyatomic molecules using Pauling's electronegativity concept. *J. Phys. Chem.*, **62**, 947–951.

50. Szwarc, M., Ghosh, B.N., and Sehon, A.H. (1950). The C-Br bond dissociation energy in benzyl bromide and allyl bromide. *J. Chem. Phys.*, **18**, 1142–1149.

51. Field, F.H. and Franklin, J.L. (1957). *Electron impact phenomena and the properties of gaseous ions*. Academic Press, New York, NY.

52. Smith, D.W. (1987). An acidity scale for binary oxides. *J. Chem. Edu.*, **64**, 480–482.

53. Brown, I.D. and Skowron, A. (1990). Electronegativity and Lewis acid strength. *J. Am. Chem. Soc.*, **112**, 3401–3403.

54. Gordy, W. and Orville Thomas, W.J. (1956). Electronegativities of the Elements. *J. Chem. Phys*, **24**, 439–444.

55. Conway, B.E. and Bockris, J.O'M, (1957). Electrolytic hydrogen evolution kinetics and its relation to the electronic and adsorptive properties of the metal. *J. Chem. Phys.*, **26**, 532–541.

56. Miedema, A.R., de Boer, F.R., and de Chatel, P.F. (1973). Empirical description of the role of electronegativity in alloy formation. *J. Phys. F: Metal Phys.*, **3**, 1558–1562.

57. Trasatti, S. (1972). Electronegativity, work function, and heat of adsorption of hydrogen on metals. *J. Chem. Soc., Faraday Trans.*, **168**, 229–236.

58. Michaelson, H.B. (1978). Relation between an atomic electronegativity scale and the work function. *IBM J. Res. Develop.*, **22**, 72–80.

59. Mulliken, R.S. (1934). A new electroaffinity scale; Together with data on valence states and on valence ionization potentials and electron affinities. *J. Chem. Phys.*, **2**, 782–791.

60. Ghosh, D.C. (2005). A new scale of electronegativity based on absolute radii of atoms. *J. Theoret. Comput. Chem.* **4**, 21–33.

61. Mulliken, R.S. (1952). Molecular compounds and their spectra. II. *J. Am. Chem. Soc.*, **74**, 811–824.

62. Ghosh, D.C. and Islam, N. (2011). Whether electronegativity and hardness are manifest two different descriptors of the one and the same fundamental property of atoms – A quest. *Int. J. Quantum Chem.*, **111**, 40–51.

63. Putz, M.V. (2008). *Absolute and chemical electronegativity and hardness*. Nova Science Publishers, New York.

64. Ayers, P.W. (2007). The physical basis of the hard/soft acid/base principle. *Faraday Discuss*, **135**, 161–190.

65. Li, K., Wang, X., Zhang, F., and Xue, D. (2008). Electronegativity identification of novel super hard materials. *Phys. Rev. Lett.*, **100**, 235504–235507.

66. Nagle, J.K. (1990). Atomic polarizability and electronegativity. *J. Am. Chem. Soc.*, **112**, 4741–4747.

67. Ghosh, D.C. and Gupta, K. (2006). A new scale of electronegativity of 54 elements of periodic table based on polarizability of atoms. *J. Theor. Comput. Chem.*, **5**, 895–911.

68. Pearson, R.G. (1963). Hard and soft acids and bases. *J. Am. Chem. Soc.*, **85**, 3533–3539.

69. Parr, R.G. and Pearson, R.G. (1983). Absolute hardness: Companion parameter to absolute electronegativity. *J. Am. Chem. Soc.*, **105**, 7512–7516.

70. Pearson, R.G. (1985). Absolute electronegativity and absolute hardness of Lewis acids and bases. *J. Am. Chem. Soc.*, **107**, 6801–6806.

71. Ahrland, S., Chatt, J., and Davies, N.R. (1958). The relative affinities of ligand atoms for acceptor molecules and ions. *Q. Reu. Chem. Sot.*, **12**, 265–276.

72. Chatt, J. (1958). The stabilisation of low valent states of the transition metals: Introductory lecture. *J. Inorg. Nurl. Chem.*, **8**, 515–531.

73. Ahrland, S. (1968). Thermodynamics of complex formation between hard and soft acceptors and donors. *Structure and Bonding*, **5**, 118–149.

74. Baird, N.C. and Whitehead, M.A. (1964). Ionic character. *Theor. Chim. Acta*, **2**, 259–264.

75. Orsky, A.R. and Whitehead, M.A. (1987). Electronegativity in density functional theory: Diatomic bond energies and hardness parameters. *Can. J. Chem.*, **65**, 1970–1979.

76. Garner-O' Neale, L.D., Bonamy, A.F., Meek, T.L., and Patrick, B.G. (2003). Calculating group electronegativities using the revised Lewis–Langmuir equation. *J. Mol. Struct.* (THEOCHEM), **639**, 151–156.

77. Mullay, J. (1984). Atomic and group electronegativities. *J. Am. Chem. Soc.*, **106**, 5842–5847.

78. KeYan, L. and DongFeng, X. (2009). New development of concept of electronegativity. *Chinese Sci. Bull.*, **54**, 328–334.

79. Spieseke, H. and Schneider, W.G. (1961). Effect of electronegativity and magnetic anisotropy of substituents on C13 and H1 chemical shifts in $CH_3X$ and $CH_3CH_2X$ compounds. *J. Chem. Phys.*, **35**, 722–731.

80. Clasen, C.A. and Good, M.L. (1970). Interpretation of the Moessbauer spectra of mixed-hexahalo complexes of tin (IV). *Inorg. Chem.*, **9**, 817–820.

81. Ichikawa, S. (1989). High-Tc superconductors and weighted harmonic mean electronegativities. *J. Phys. Chem.*, **93**, 7302–7311.

82. Devautour, S., Giuntini, J.C., Henn, F., Douillard, J.M., Zanchetta, J.V., and Vanderschueren, J. (1999). Application of the Electronegativity Equalization Method to the Interpretation of TSDC Results: Case of a Mordenite Exchanged by $Na^+$ and $Li^+$ Cations. *J. Phys. Chem. B*, **103**, 3275–3281.

83. Schaeffera, J.K., Gilmera, D.C., Capassoa, C., Kalpata, S., Taylora, B., Raymonda, M.V., Triyosoa, D., Hegdea, R., Samavedama, S.B., and White, Jr, B.E. (2007). Application of group electronegativity concepts to the effective work functions of metal gate electrodes on high-[kappa] gate oxides. *Microelectronic Engineering*, **84**, 2196–2203.

84. Baeten, A. and Geerlings, P. (1999). The use of the electronegativity equalization principle to study charge distributions in enzymes: Application to dipeptides. *J. Mol. Struct.* (THEOCHEM), **465**, 203–207.

85. Ramsden, C.A. (2004). The influence of electronegativity on triangular three-centre two-electron bonds: The relative stability of carbonium ions, [pi]-complex chemistry and the-2h [beta] effect. *Tetrahedron*, **60**, 3293–3309.

86. Reddy, R.R., Gopal, K.R., Narasimhulu, K., Reddy, L.S.S., Kumar, K.R., Reddy, C.V.K., and Ahmed, S.N. (2008). Correlation between optical electronegativity and refractive index of ternary chalcopyrites, semiconductors, insulators, oxides and alkali halides. *Opt. Mat.*, **31**, 209–212.

87. Douillard, J.M., Salles, F., Henry, M., Malandrini, H., and Clauss, F. (2007). Surface energy of talc and chlorite: Comparison between electronegativity calculation and immersion results. *J. Col. Int. Sci.*, **305**, 352–360.

88. Kwon, C.W., Poquet, A., Mornet, S., Campet, G., Delville, M.H., Treguer, M., and Portier, J. (2001). Electronegativity and chemical hardness: Two helpful concepts for understanding oxide nanochemistry. *Materials Letters*, **51**, 402–413.

89. Makino, Y. (1994). Structural mapping of binary compounds between transitional metals on the basis of bond orbital model and orbital electronegativity. *Intermetallics*, **2**, 67–72.

90. Ray, L.F., Kristy, L.E., Weier, M.L., McKinnon, A.R., Williams, P.A., and Leverett, P. (2005). Thermal decomposition of agardites (REE)—Relationship between dehydroxylation temperature and electronegativity. *Therm. Acta*, **427**, 167–170.

91. Tang, J.X., Lee, C.S., Lee, S.T., and Xu, Y.B. (2004). Electronegativity and charge-injection barrier at organic/metal interfaces. *Chem. Phys. Lett.*, **396**, 92–96.

92. Portier, J., Campet, G., Etourneau, J., Shastry, M.C.R., and Tanguy, B. (1994). A simple approach to materials design: Role played by an ionic-covalent parameter based on polarizing power and electronegativity. *J. Alloys. Compd.*, **209**, 59–64.

93. Mortier, W.J., Van Genechten, K., and Gasteiger, J. (1985). Electronegativity equalization: Application and parametrization. *J. Am. Chem. Soc.*, **107**, 829–835.

94. Mortier, W.J., Ghosh, S.K., and Shankar, S. (1986). Electronegativity-equalization method for the calculation of atomic charges in molecules. *J. Am. Chem. Soc.*, **108**, 4315–4320.

95. Van Genechten, K.A., Mortier, W.J., and Greelings, P. (1987). Intrinsic framework electronegativity: A novel concept in solid state chemistry. *J. Chem. Phys.*, **86**, 5063–5071.

96. Svobodová Vařeková, R, Jiroušková, Z., Vaněk, J., Suchomel, Š., and Kočalnt, J. (2007). Electronegativity equalization method: Parameterization and validation for large sets of organic, organohalogene and organometal molecule. *Int. J. Mol. Sci.*, **8**, 572–582.

97. Chandra, A.K. and Nguyen, M.T. (2002). Use of local softness for the interpretation of reaction mechanisms. *Int. J. Mol. Sci.*, **3**, 310–323.

98. Parr, R.G. and Yang, W. (1984). Density functional approach to the frontier-electron theory of chemical reactivity. *J. Am. Chem. Soc.*, **106**, 4049–4050.

99. Parr, R.G., Szentpály, L.V., and Liu, S. (1999). Electrophilicity index. *J. Am. Chem. Soc.*, **121**, 1922–1924.

100. Chattaraj, P.K. Sarkar, U., and Roy, D.R. (2006). Electrophilicity index. *Chem. Rev.*, **106**, 2065–2091.

101. Cohen, M.H. and Wasserman, A. (2006). On hardness and electronegativity equalization in chemical reactivity theory. *J. Stat. Phys.*, **125**, 1121–1139.

102. Berkowitz, M. and Parr, R.G. (1988). Molecular hardness and softness, local hardness and softness, hardness and softness kernels, and relations among these quantities. *J. Chem. Phys.*, **88**, 2554–2557.

103. Hannay, N.B. and Smyth, C.P. (1946). The dipole moment of hydrogen fluoride and the ionic character of bonds. *J. Am. Chem. Soc.*, **68**, 171–173.

104. Sanderson, R.T. (1988). Principles of electronegativity Part I. General Nature. *Chem. Educ.*, **65**(2), 112–118.

105. Sanderson, R.T. (1988). Principles of electronegativity Part II. Applications. *J. Chem. Educ.*, **65**(3), 227–231.

# 3

1. Avella, M., De Vlieger, J.J., Errico, M.E., Fischer, S., Vacca, P., and Volpe, M.G. (2005). Biodegradable starch/clay nanocomposite films for food packaging applications. *Food Chem.*, **93**(3), 467–474.

2. Thakore, I.M., Desai, S., Sarawade, B.D., and Devi, S. (2001). Studies on biodegradability, morphology and thermomechanical properties of LDPE/modified starch blends. *Eur. Polym. J.*, **37**(1), 151–160.

3. Chiu, F.C., Lai, S.M., and Ti, K.T. (2009). Characterization and comparison of metallocene-catalyzed polyethylene/thermoplastic starch blends and nanocomposites. *Polym. Test.*, **28**(3), 243–250.

4. Huang, C.Y., Roan, M.L., Kuo, M.C., and Lu, W.L. (2005). Effect of compatibiliser on the biodegradation and mechanical properties of high-content starch/low-density polyethylene blends. *Polym. Degradation. Stab.*, **90**(1), 95–105.

5. Tang, S., Zou, P., Xiong, H., and Tang, H. (2008). Effect of nano-SiO2 on the performance of starch/polyvinyl alcohol blend films. *Carbohyd. Polym.*, **72**(3), 521–526.

6. Ma, X., Yu, J., and Wang, N. (2007). Production of thermoplastic starch/MMT-sorbitol nanocomposites by dual-melt extrusion processing. *Macromol. Mater. Eng.*, **292**(6), 723–728.

7. Park, H.R., Chough, S.H., Yun, Y.H., and Yoon, S.D. (2005). Properties of starch/PVA blend films containing citric acid as additive. *J. Polym. Environ.*, **13**(4), 375–382.

8. Carvalho, A.J.F., Job, A.E., Curvelo, A.A.S., and Gandini, A. (2003). Thermoplastic starch/natural rubber blends. *Carbohyd. Polym.*, **53**(1), 1–5.

9. Mano, J.F., Koniarova, D., and Reis, R.L. (2003). Thermal properties of thermoplastic starch/synthetic polymer blends with potential biomedical applicability. *J. Mater. Sci. Mater. Med.*, **14**(2), 127–135.

10. Choudalakis, G. and Gotsis, A.D. (2009). Permeability of polymer/clay nanocomposites: A review. *Eur. Polym. J.*, **45**(4), 967–984.

11. Yang, J.H., Yu, J.G., and Ma, X.F. (2006). Preparation and properties of ethylenebisformamide plasticized potato starch (EPTPS). *Carbohyd. Polym.*, **63**(2), 218–223.

12. Mingliang, G. (2009). Preparation and properties of polypropylene/clay nanocomposites using an organoclay modified through solid state method. *J. Reinf. Plast. Compos.*, **28**(1), 5–15.

13. Fedullo, N., Sorlier, E., Sclavons, M., Bailly, C., Lefebvre, J.M., and Devaux, J. (2006). Polymer based nanocomposites: Overview, applications and perspectives. *Prog. Org. Coatings.*, **58**(2–3), 87–95.

14. Giannakas, A., Spanos, C.G., Koirkoumelis, N., Vaimakis, T., and Ladavos, A. (2008). Preparation, characterization and water barrier properties of PS/organo-montmorillonite nanocomposites. *Eur. Polym. J.*, **44**(12), 3915–3921.

15. Chandra, R. and Rustgi, R. (1998). *Prog. Polym. Sci., Jpn.*, **23**, 1273.

16. Ikada, Y. and Tsuji, H. (2000). *Macromol. Rapid Commun.*, **21**, 117.

17. Agrawal, C.M. and Ray, R.B. (2001). *J. Biomed. Mater. Res.*, **55**, 141.

18. Winzenburg, G., Schmidt, C., Fuchs, S., and Kissel, T. (2004). *Adv. Drug Deliv. Rev.*, **56**, 1453.

19. Koenig, M.F. and Huang, S.J. (1995). *Polymer*, **36**, 1877.

20. Bastioli, C., Cerutti, A., Guanella, I., Romano, G.C., and Tosin, M.J. (1995). *Environ. Polym. Degrad.*, **3**, 81.

21. Bastioli, C. (1998). *Polym. Degrad. Stab.*, **59**, 263.

22. Vikman, M., Hulleman, H.D., van der Zee, M., Myllarinen, H., and Feil, H.J. (1999). *Appl. Polym. Sci.*, **74**, 2594.

23. Averous, L., Moro, L., Dole, P., and Fringant, C. (2000). *Polymer*, **41**, 4157.

24. Yavuz, H. and Baba, C.J. (2003). *Polym. Environ.*, **11**, 107.

25. Singh, R.C., Pandey, J.K., Rutot, D., Degee, Ph., and Dubois, Ph. (2003). *Carbohydr. Res.*, **338**, 1759.

26. Azevedo, H.S., Gama, F.M., and Reis, R.L. (2003). *Biomacromolecules*, **4**, 1703.

27. Mano, J.F., Koniarova, D., and Reis, R.L. (2003). *J. Mater. Sci.: Mater. Med.*, **14**, 127.

28. Kweon, D.K., Kawasaki, N., Nakayama, A., and Aiba1, S. (2004). *J. Appl. Polym. Sci.*, **92**, 1716.

29. Mano, J.F., Sousa, R.A., Boesel, L.F., Neves, N.M., and Reis, R.L. (2004). *Compos. Sci. Technol.*, **64**, 789.

30. Gomes, M.E., Sikavitsas, V.I., Behravesh, H., Reis, R.L., and Mikos, A.G. (2003). *J. Biomed. Mater. Res.*, **67A**, 87.

31. Oliveira, A.L. and Reis, R.L. (2004). *J. Mater. Sci.: Mater. Med.*, **15**, 533.

32. Costa, S. A. and Reis, R.L. (2004). *J. Mater. Sci.: Mater. Med.*, **15**, 335.

33. Alves, C.M., Reis, R.L., and Hunt, J.A. (2003). *J. Mater. Sci.: Mater. Med.*, **14**, 157.

34. Marques, A. P., Reis, R.L., and Hunt, J.A. (2003). *J. Mater. Sci.: Mater. Med.*, **14**, 167.

35. Skoglund, P. and Fransson, A.J. (1996). *Appl. Polym. Sci.*, **61**, 2455.

36. Cyras, V.P., Kenny, J.M., and Vazquez, A. (2001). *Polym. Eng. Sci.*, **41**, 1521.

37. Ruseckaite, R.A., Stefani, P.M., Cyras, V.P., Kenny, J.M., and Vazquez, A.J. (2001). *Appl. Polym. Sci.*, **82**, 3275.

38. McCrum, N.G., Read, B.E., and Williams, G. (1991). *Anelastic and dielectric effects in polymer solids*. Dover, New York.

39. Ferry, J.D. (1980). *Viscoelastic properties of polymers*, 3rd ed. Wiley, New York.

40. Mano, J.F., Reis, R.L., and Cunha, A.M. (2002). Dynamic mechanical analysis in polymers for medical applications. In: *Polymer based systems on tissue engineering, replacement and regeneration*, R.L. Reis and D. Cohn (Eds.), NATO Sci. Ser., Kluwer, Dordrecht.

41. Mani, R. and Bhattacharya, M. (2001). *Eur. Polym. J.*, **37**, 515.

42. Hohne, G.W.H., Schawe, J., and Schick, J. (1993). *Thermochim. Acta*, **221**, 129.

43. Crescenzi, V., Manzini, G., Calzolari, G., and Borri, C. (1972). *Eur. Polym.J.*, **8**, 449.

44. Jang, J. and Lee, D.K. (2003). *Polymer*, **44**, 8139.

45. Cebe, J. and Hong, S.D. (1986). *Polymer*, **27**, 1183.

46. Ke, T. and Sun, X.J. (2003). *Appl. Polym. Sci.*, **89**, 1203.

47. Dobreva-Veleva, A. and Gutzow, I.J. (1993). *Non-Cryst. Solids*, **162**, 13.

48. Alonso, M., Velasco, J.I., and de Saja, J.A. (1997). *Eur. Polym. J.*, **33**, 255.

49. Díez-Gutiérrez, S., Rodríguez-Pérez, M.A., de Saja, J.A., and Velasco, J.I. (1999). *Polymer*, **40**, 5345.

50. Velasco, J.I., Morhain, C., Martínez, A.B., Rodríguez-Pérez, M.A., and de Saja, J.A. (2002). *Polymer*, **43**, 6813.

51. Ozawa, T. (1971). *Polymer*, **12**, 150.

52. Liu, T.X., Mo, Z.S., Wang, S.E., and Zhang, H.F. (1997). *Polym. Eng. Sci.*, **37**, 568.

53. Di Lorenzo, M.L. and Silvestre, C. (1999). *Prog. Polym. Sci.*, **24**, 917.

54. Wang, Y., Shen, C., Li, H., and Li, Q.J. Chen, J. (2004). *Appl. Polym. Sci.*, **91**, 308.

55. Wang, Y., Shen, C., and Chen, J. (2003). *Polym. J. (Tokyo)*, **35**, 884.

56. Kim, S.H., Ahn, S.H., and Hirai, T. (2003). *Polymer*, **44**, 5625.

57. Day, M., Cooney, J.D., Shaw, K., and Watts, J. (1998). *Therm. Anal.Calorim.*, **52**, 261.

58. Angell, C.A. (1991). *J. Non-Cryst. Solids.*, **13**, 131–133.

59. Bohmer, R., Ngai, K.L., Angell, C.L., and Plazek, D.J. (1993). *J. Chem. Phys.*, **99**, 4201.

60. Buechner, P.M., Lakes, R.S., Swan, C., and Brand, C.A. (2001). *Ann.Biomed. Eng.*, **29**, 719.

61. Weeraphat, P., Siwaporn, M., Ming, T.I. (2008). Formation of hydroxyapatite crystallites using organic template of polyvinyl alcohol (PVA) and sodium dodecyl sulfate (SDS). *Mat. Chem. Phys.*, **112**, 453–460.

62. Ramakrishna, S., Mayer, J., Winter, E., Kam, M., and Leong, W. (2001). Biomedical applications of polymer-composite materials: a review. *Compos. Sci. Technol.*, **67**, 1189–1224.

63. Kikuchi, M., Itoh, S., Ichinose, S., Shinomiya, K., and Tanaka, J. (2001). Self-organization mechanism in a bone-like hydroxyapatite/collagen nanocomposite synthesized in vitro and its biological reaction *in vivo*. *Biomaterials*, **22**, 1705–1711.

64. Toworfe, G.K., Composto, R.J., Shapiro, I.M., and Ducheyne, P. (2006). Nucleation and growth of calcium phosphate on amine, carboxyl- and hydroxyl-silane self-assembled monolayers. *Biomaterials*, **27**, 631–642.

65. Mollazadeh, S., Javadpour, J., and Khavandi, A. (2007). In situ synthesis and characterization of nano-size hydroxyapatite in poly(vinyl alcohol) matrix. *Ceramics International*, **33**, 1579–1583.

66. Alexeev, V.L., Kelberg, E.A., and Evmenenko, G.A. (2000). Improvement of the mechanical properties of chitosan films by the addition of poly (ethylene oxide). *Polym. Eng. Sci.*, **40**, 1211–1215.

67. Clavaguera, N., Saurina, J., Lheritier, J., Masse, J., Chauvet, A., and Mora, M.T. (1997). Eutectic mixtures for pharmaceutical applications: A thermodynamic and

kinetic study. *Thermochim. Acta.*, **290**, 173–180.

68. Zhang, S. and Gonsalves, K.E. (1997). Preparation and characterization of thermally stable nanohydroxyapatite. *J. Mater. Sci. Mater. Med.*, **8**, 25–28.

69. Calvert, P. and Rieke, P. (1996). Biomimetic mineralization in and on polymers. *Chem. Mater*, **8**, 1715–1727.

70. Leonor, I.B., Baran, E.T., Kawashita, M., Reis, R.L., Kokubo, T., and Nakamura, T. (2008). Growth of a bonelike apatite on chitosan microparticles after a calcium silicate treatment. *Acta Biomaterialia*, **4**, 1349–1359.

71. Perren, S.M. and Gogolewski. (1994). *Clinical requirements for bioresorbable implants internal fixation*, World Scientific, Hong Kong. **11**, 35–43.

72. Katti, K. and Gujjula, P. (2002). Control of mechanical responses in in-situ polymer/hydroxyapatite composite for bone replacement. *Proceedings of the15th ASCE Engineering Mechanism Conference*, 2–5 June, Columbia University, New York, NY.

73. Ciftcioglu, N. and Mckay, D.S. (2005). Overeiew of biominralization and nanobacteria. *Lunar and Planetary Science*, **VI**, 1205–1215.

74. Mano, J.F., Sousa, R.A., Boesel, L.F., Neves, N.M., and Reis, R.L. (2004). Bioinert biodegradable and injectable polymeric matrix composites for hard tissue replacement: state of the art and recent developments. *Compos. Sci. Technol.*, **64**, 789–817.

75. Galliard, T. (1987). *Starch availability and utilization* (Vol. 6, pp. 1–15), John Wiley & Sons, London.

76. Kokubo, T., Kim, H.M., and Kawashata, M. (2004), Novel bioactive materials with different mechanical properties. *Biomaterials*, **24**, 2161–2175.

77. Dalas, E. and Chrissanthopoulos, A. (2003). The overgrowth of hydroxyapatite on new functionalized polymers. *J. Cryst. Growth*, **255**, 163–169.

78. Chen, J., Wei, K., Zhang, S.H., and Wang, X. (2006). Surfactant-assisted synthesis of hydroxyapatite particles. *Materials Letters*, **60**, 3227–3231.

79. Sinha, A. and Guha, A. (2008). Biomimetic patterning of polymer hydrogels with hydroxyapatite nanoparticles. *Mater. Sci. Eng. C.*, **29**, 1330–1333.

80. Wang, L. and Chunzhong, L. (2007). Preparation and physicochemical properties of a novel hydroxyapatite/chitosan–silk fibroin composite. *Carb. Polym.*, **68**, 740–745.

81. Ahmad, M.B., Shameli, K., Darroudi, M., Yunus, W.M.Z.W., and Ibrahim, N.A. (2009a). Synthesis andCharacterization of Silver/Clay/Chitosan Bionanocomposites by UV-Irradiation Method. *Am. J. Appl. Sci.*, **6**(12), 2030–2035.

82. Ahmad, M.B., Shameli, K., Darroudi, M., Yunus, W.M.Z.W., and Ibrahim, N.A. (2009b). Synthesis and Characterization of Silver/Clay Nanocomposites by Chemical Reduction Method. *Am. J. Appl. Sci.*, **6**(11), 1909–1914.

83. Alemdar, A., Güngör, N., Ece, Ö.I., and Atici, O. (2005). The rheological properties and characterization of bentonite dispersions in the presence of non-ionic polymer PEG. *J. Mater. Sci.*, **40**, 171–177.

84. Belova, V., Möhwald, H., and Shchukin, D.G. (2008). Sonochemical intercalation of preformed gold nanoparticles into multilayered clays. *Langmuir*, **24**, 9747–9753.

85. Chen, P., Song, L. Liu, Y., and Fang, Y. (2007). Synthesis of silver nanoparticles by [gamma]-ray irradiation in acetic water solution containing chitosan. *Radiat. Phys. Chem.*, **76**, 1165–1168.

86. Darder, M., Aranda, P., and Ruiz-Hitzky, E. (2007). Bionanocomposites: A new concept of ecological, bioinspired, and functional hybrid materials. *Adv. Mat.*, **19**, 1309–1319.

87. Darroudi, M., Ahmad, M.B. Shameli, K. Abdullah, A.H., and Ibrahim, N.A. (2009). Synthesis and characterization of UV-irradiated silver/montmorillonite nanocomposites. *Solid State Sci.*, **11**, 1621–1624.

88. Fang, J.M., Fowler, P.A., Sayers, C., and Williams, P. A. (2004). The chemical modification of a range of starches under aqueous reaction condition. *Carbohyd Polym.*, **55**, 283–289.

89. Hongshui, W., Qiao, X., Chen, J., and Ding, S. (2005). Preparation of silver nanoparticles by chemical reduction method.

*Colloids and Surfaces A: Physicochem. Eng. Aspects*, **256**, 111–115.

90. Huang, M.F., Yu, J.G., and. Ma, X.F. (2006). High mechanical performance MMT-urea and Formamideplasticized thermoplastic cornstarch biodegradable nanocomposites. *Carbohyd. Polym.*, **63**, 393–399.

91. Kampeerapappun, P., Aht-Ong, D., Pentrakoon, D., and Srikulkit, K. (2007). Preparation of cassava starch/montmorillonite composite film. *Carbohyd. Polym.*, **67**, 155–163.

92. Khanna, P.K., Singh, N., Charan, S., Subbarao, V.V.V.S., Gokhale, R., and Mulik, U.P. (2005). Synthesis and characterization of Ag/PVA nanocomposite by chemical reduction method. *Mater. Chem. Phys.*, **93**, 117–121.

93. Kozak, M. and Domka, L.J. (2004). Adsorption of the quaternary ammonium salts on montmorillonite. *J.Phys. Chem. Solids.*, **65**, 441–445.

94. Mangiacapra, P., Gorrasi, G., Sorrentino, A., and Vittoria, V. (2006). Biodegradable nanocomposites obtained by ball milling of pectin and montmorillonites. *Carbohyd. Polym.*, **64**, 516–523.

95. Mano, J.F., Koniarova, D., and Reis, R.L. (2003). Thermal properties of thermoplastic starch/synthetic polymer blends with potential biomedical applicability. *J. Mater. Sci.: Mater. Med.*, **14**, 127–135.

96. Patakfalvi, R., Oszko, A., and Dékány, I. (2003). Synthesis and characterization of silver nanoparticle/kaolinite composites. *Coll. Surf. A.*, **220**, 45–54.

97. Prasad, V., Souza, C.D., Yadav, D., Shaikh, A.J., and Vigneshwaran, N. (2006). Spectroscopic characterization of zinc oxide nanorods synthesized by solid-state reaction. *Spectrochim. Acta A.*, **65**, 173–178.

98. Raveendran, P., Fu, J., and Wallen, S. L. (2003). Completely "Green" synthesis and stabilization of metal nanoparticles. *J. Am. Chem. Soc.*, **125**, 13940–13941.

99. Temgire, M.K. and Joshi, S.S. (2004). Optical and structural studies of silver nanoparticles. *Radiat. Phys.Chem.*, **71**, 1039–1044.

100. Twu, Y.K., Chen, Y.W., and Shih, C.M. (2008). Preparation of silver nanoparticles using chitosan suspensions. *Powder Technol.*, **185**, 251–257.

101. Yin, H., Yamamoto, T., Wada, Y., and Yanagida, S. (2004). Large-scale and size-controlled synthesis of silver nanoparticles under microwave irradiation. *Mater. Chem. Phys.*, **83**, 66–70.

102. Zhao, X.P., Wang, B.X., and Li, J. (2008). Synthesis and electrorheological activity of a modified kaolinite/carboxymethyl starch hybrid nanocomposite. *J. Appl. Polym. Sci.*, **108**, 2833–2839.

103. Yu, L., Dean, K., and Li, L. (2006). Polymer blends and composites from renewable resources. *Prog. Polym. Sci.*, **31**, 576–602.

104. de Menezes, A.J., Pasquini, D., Curvelo, A.A.S., and Gandini, A. (2007). Novel thermoplastic materials based on the outer-shell oxypropylation of corn starch granules. *Biomacromolecules*, **8**, 2047–2050.

105. Mohanty, A.K., Misra, M., and Hinrichsen, G. (2000). Biofibres, biodegradable polymers and biocomposites: An overview. *Macromol. Mater. Eng.*, **276–277**, 1–24.

106. Mathew, A.P. and Dufresne, A. (2002). Morphological investigation of nanocomposites from sorbitol plasticized starch and tunicin whiskers. *Biomacromolecules*, **3**, 609–617.

107. Tábi, T. and Kovács, J.G. (2007). Examination of injection moulded thermoplastic maize starch. *Express Polym. Lett.*, **1**, 804–809.

108. Santayanon, R. and Wootthikanokkhan, J. (2003). Modification of cassava starch by using propionic anhydride and properties of the starch-blended polyester polyurethane. *Carbohydr. Polym.*, **51**, 17–24.

108. Cao, X., Chang, P.R., and Huneault, M.A. (2008). Preparation and properties of plasticized starch modified with poly(ε-caprolactone) based waterborne polyurethane. *Carbohydr. Polym.*, **71**, 119–125.

109. Cao, X., Zhang, L., Huang, J., Yang, G., and Wang, Y. (2003). Structure-properties relationship of starch/waterborne polyurethane composites. *J. Appl. Polym. Sci.*, **90**, 3325–3332.

110. Chen, Y., Cao, X., Chang, P.R., and Huneault, M.A. (2008). Comparative study on the films of poly(vinyl alcohol)/pea starch

nanocrystals and poly(vinyl alcohol)/native pea starch. *Carbohydr. Polym.*, **73**, 8–17.

111. Cao, X. and Zhang, L. (2005): Miscibility and properties of polyurethane/benzyl starch semi-interpenetrating polymer networks. *J. Polymer Sci., Part B: Polymer Phys.*, **43**, 603–615.

112. Cao, X. and Zhang, L. (2005). Effects of molecular weight on the miscibility and properties of polyurethane/benzyl starch semi-interpenetrating polymer networks. *Biomacromolecules*, **6**, 671–677.

113. Cao, X., Wang, Y., and Zhang L. (2005). Effects of ethyl and benzyl groups on the miscibility and properties of castor oil-based polyurethane/starch derivative semi-interpenetrating polymer networks. *Macromol. Biosci.*, **5**, 863–871.

114. Suda, K., Kanlaya, M., and Manit, S. (2002). Synthesis and property characterization of cassava starch grafted poly[acrylamide-co-(maleic acid)] superabsorbent via γ-irradiation. *Polymer*, **43**, 3915–3924.

115. Lepifre, S., Froment, M., Cazaux, F., Houot, S., Lourdin, D., Coqueret, X., Lapierre, C., and Baumberger, S. (2004). Lignin incorporation combined with electron-beam irradiation improves the surface water resistance of starch films. *Biomacromolecules*, **5**, 1678–1686.

116. Chen, B. and Evans, J.R.G. (2005). Thermoplastic starch-clay nanocomposites and their characteristics. *Carbohydr. Polym.*, **61**, 455–463.

117. Cao, X., Chen, Y., Chang, P.R., and Huneault, M.A. (2007). Preparation and properties of plasticized starch/multiwalled carbon nanotubes composites. *J. Appl. Polym. Sci.*, **106**, 1431–1437.

118. Šturcová A., Davies G.R., and Eichhorn, S.J. (2005) Elastic modulus and stress-transfer properties of tunicate cellulose whiskers. *Biomacromolecules*, **6**, 1055–1061.

119. Azizi Samir M.A.S., Alloin, F., and Dufresne, A. (2005). Review of recent research into cellulosic whiskers, their properties and their application in nanocomposite field. *Biomacromolecules*, **6**, 612–626.

120. Dubief, D., Samain, E., and Dufresne, A. (1999). Polysaccharide microcrystals reinforced amorphous poly(β-hydroxyoctanoate) nanocomposite materials. *Macromolecules*, **32**, 5765–5771.

121. Wang, Y., Cao, X., and Zhang, L. (2006). Effects of cellulose whiskers on properties of soy protein thermoplastics. *Macromol. Biosci.*, **6**, 524–531.

122. Noshiki, Y., Nishiyama, Y., Wada, M., Kuga, S., and Magoshi, J. (2002). Mechanical properties of silk fibroinmicrocrystalline cellulose composite films. *J. Appl. Polym. Sci.*, **86**, 3425–3429.

123. Helbert, W., Cavaille, J.Y., and Dufresne, A. (1996). Thermoplastic nanocomposites filled with wheat straw cellulose whiskers. Part I: Processing and mechanical behavior. *Polym. Compos.*, **17**, 604–611.

# 4

1. Ziabari, M., Mottaghitalab, V., and Haghi, A.K. (2008). *Korean J. Chem. Eng.*, **25**, 923.

2. Haghi, A.K. and Akbari, M. (2007). *Phys. Stat. Sol. A.*, **204**, 1830.

3. Kanafchian, M., Valizadeh, M., and Haghi, A.K. (2011). *Korean J. Chem. Eng.*, **28**, 428.

4. Ziabari, M., Mottaghitalab, V., and Haghi, A.K. (2008). *Korean J. Chem. Eng.*, **25**, 905.

5. Kanafchian, M., Valizadeh, M. and Haghi, A.K. (2011). *Korean J. Chem. Eng.*, **28**, 445.

6. Lee, S. and Obendorf, S.K. (2006). *J. Appl. Polym. Sci.*, 102, 3430.

7. Lee, S., Kimura, D., Lee, K.H., Park, J.C., and Kim, I.S., (2010). *Textile Res. J.*, **80**, 99.

8. Pedicini, A. and Farris, R.J. (2003). *Polymer* **44**, 6857.

9. Lee, K.H., Lee, B.S., Kim, C.H., Kim, H.Y., Kim, K.W., and Nah, C.W. (2005). *Macromol. Res.*, **13**, 441.

10. Lee, S.M., Kimura, D., Yokoyama, A., Lee, K.H., Park, J.C., and Kim, I.S. (2009). *Textile Res. J.*, **79**, 1085.

11. Liu, L., Huang, Z.M., He, C.L., and Han, X.J. (2006). *Mater. Sci. Eng. A.*, **435–436**, 309.

12. Fung, W. (2002). Materials and their properties. In: *Coated and laminated textiles*, (1st ed., pp. 63–71). Woodhead Publishing.

# 5

1. Agarwal, S., Wendorff, J.H., and Greiner, A. (2008). *Polymer*, **49**, 5603.

2. Li, M., Mondrinos, M.J., Gandhi, M.R., Ko, F.K., Weiss, A.S., and Lelkes, P.I. (2005). *Biomaterials*, **26**, 5999.

3. Zeng, J., Yang, L., Liang, Q., Zhang, X., Guan, H., Xu, X., Chen, X., and Jing, X. (2005). *J. Control. Release.*, **105**, 43.

4. Khil, M.-S., Cha, D.-I., Kim, H.-Y., Kim, I.-S., and Bhattarai, N. (2003). *J. Biomed. Mat. Res. B*, **67B**, 675.

5. Taylor, G.I. (1969). *Proc. Roy Soc. London*, **313**, 453.

6. Doshi, J. and Reneker, D.H. (1995). *J. Electrostat.*, **35**, 151.

7. Li, D. and Xia, Y. (2004). *Adv. Mater.*, **16**, 1151.

8. Ziabari, M., Mottaghitalab, V., and Haghi, A.K. (2008). *Korean J. Chem. Eng.*, **25**, 923.

9. Tan, S.H., Inai, R., Kotaki, M., and Ramakrishna, S. (2005). *Polymer*, **46**, 6128.

10. Sukigara, S., Gandhi, M., Ayutsede, J., Micklus, M., and Ko, F. (2003). *Polymer*, **44**, 5721.

11. Matthews, J.A., Wnek, G.E., Simpson, D.G., and Bowlin, G.L. (2002). *Biomacromolecules*, **3**, 232.

12. McManus, M.C., Boland, E.D., Simpson, D.G., Barnes, C.P., and Bowlin, G.L. (2007). *J. Biomed. Mater. Res. A*, **81A**, 299.

13. Huang, Z.-M., Zhang, Y.Z., Ramakrishna, S., and Lim, C.T. (2004). *Polymer*, **45**, 5361.

14. Zhang, X., Reagan, M.R., and Kaplan, D.L. (2009). *Adv. Drug. Deliver. Rev.*, **61**, 988.

15. Noh, H.K., Lee, S.W., Kim, J.-M., Oh, J.-E., Kim, K.-H., Chung, C.-P., Choi, S.-C., Park, W.H., and Min, B.-M. (2006). *Biomaterials*, **27**, 3934.

16. Ohkawa, K., Minato, K.-I., Kumagai, G., Hayashi, S., and Yamamoto, H. (2006). *Biomacromolecules*, **7**, 3291.

17. Agboh, O.C. and Qin, Y. (1997). *Polym. Adv. Technol.*, **8**, 355.

18. Rinaudo, M. (2006). *Prog. Polym. Sci.*, **31**, 603.

19. Aranaz, I., Mengíbar, M., Harris, R., Paños, I., Miralles, B., Acosta, N., Galed, G., and Heras, Á. (2009). *Curr. Chem. Biol.*, **3**, 203.

20. Neamnark, A., Rujiravanit, R., and Supaphol, P. (2006). *Carbohydr. Polym.*, **66**, 298.

21. Duan, B., Dong, C., Yuan, X., and Yao, K. (2004). *J. Biomater. Sci. Polymer. Ed.*, **15**, 797.

22. Jia, Y.-T., Gong, J., Gu, X.-H., Kim, H.-Y., Dong, J., and Shen, X.-Y. (2007). *Carbohydr. Polym.*, **67**, 403.

23. Homayoni, H., Ravandi, S.A.H., and Valizadeh, M. (2009). *Carbohydr. Polym.*, **77**, 656.

24. Geng, X., Kwon, O.-H., and Jang, J. (2005). *Biomaterials*, **26**, 5427.

25. Torres-Giner, S., Ocio, M.J., and Lagaron, J.M. (2008). *Anglais*, **8**, 303.

26. Vrieze, S.D., Westbroek, P., Camp, T.V., and Langenhove, L.V. (2007). *J. Mater. Sci.*, **42**, 8029.

27. Ohkawa, K., Cha, D., Kim, H., Nishida, A., and Yamamoto, H. (2004). *Macromol. Rapid Comm.*, **25**, 1600.

28. Iijima, S. (1991). *Nature*, **354**, 56.

29. Esawi, A.M.K. and Farag, M.M. (2007). *Mater. Design.*, **28**, 2394.

30. Feng, W., Wu, Z., Li, Y., Feng, Y., and Yuan, X. (2008). *Nanotechnology*, **19**, 105707.

31. Liao, H., Qi, R., Shen, M., Cao, X., Guo, R., Zhang, Y., and Shi, X. (2011). *Colloid. Surface B.*, DOI: 10.1016/j.colsurfb.2011.02.010

32. Baek, S.-H., Kim, B., and Suh, K.-D. (2008). *Colloid. Surface A*, **316**, 292.

33. Liu, Y.-L., Chen, W.-H., and Chang, Y.-H. (2009). *Carbohydr. Polym.*, **76**, 232.

34. Tkac, J., Whittaker, J.W., and Ruzgas, T. (2007). *Biosens. Bioelectron.*, **22**, 1820.

35. Spinks, G.M., Geoffrey, M., Shin, S.R., Wallace, G.G., Whitten, P.G., Kim, S.I., and Kim, S.J. (2006). *Sensor. Actuat. B-Chem.*, **115**, 678.

36. Zhang, H., Wang, Z., Zhang, Z., Wu, J., Zhang, J., and He, J. (2007). *Adv. Mater.*, **19**, 698.

37. Deitzel, J.M., Kleinmeyer, J., Harris, D., and Beck Tan, N.C. (2001). *Polymer*, **42**, 261.

38. Zhang, S., Shim, W.S., and Kim, J. (2009). *Mater. Design*, **30**, 3659.

39. Li, Y., Huang, Z., and Lu, Y. (2006). *Eur. Polym. J.*, **42**, 1696.

# 6

1. Liang, C., Lingling, X., Hongbo, S., and Zhibin, Z. (2009). *J. Energy Convers. Manage.*, **50**, 723.

2. Sharma, A., Tyagi, V.V., Chen, C.R., and Buddhi, D. (2009). *J. Renewable Sustainable Energy Rev.*, **13**, 318.

3. Regin, A.F., Solanki, S.C., and Saini, J.S. (2008). *J. Renewable Sustainable Energy Rev.*, **12**, 2438.

4. Meng, Q. and Hu, J. (2008). *J. Sol. Energy Mater. Sol. Cells*, **92**, 1260.

5. Mehling, H. and Cabeza, L.F. (2008). *Heat and cold storage with PCM*. Springer, Berlin.

6. Mondal, S. (2008). *J. Appl. Therm. Eng.*, **28**, 1536.

7. Onder, E., Sarier, N., and Cimen, E. (2008). *J. Thermochim. Acta*, **467**, 63.

8. Xing, L., Hongyan, L., Shujun, W., Lu, Z., and Hua, C. (2006). *J. Sol. Energy*, **80**, 1561.

9. Alkan, C., Sari, A., Karaipekli, A., and Uzun, O. (2009). *J. Sol. Energy Mater. Sol. Cells*, **93**, 143.

10. Fang, G., Li, H., Yang, F., Liu, X., and Wu, S. (2009). *Chem. Eng. J.*, **153**, 217.

11. Fang, Y., Kuang, S., Gao, X., and Zhang, Z. (2008). *J. Energy Convers. Manage.*, **49**, 3704.

12. Chen, C., Wang, L., and Huang, Y. (2008). *J. Mater. Lett.*, **62**, 3515.

13. Chen, C., Wang, L., and Huang, Y. (2009). *J. Chem. Eng.*, **150**, 269.

14. Chen, C., Wang, L., and Huang, Y. (2007). *Polymer com.*, **48**, 5202.

15. Chen, C., Wang, L., and Huang, Y. (2008). *J. Sol. Energy Mater. Sol. Cells*, **92**, 1382.

16. Chen, C., Wang, L., and Huang, Y. (2009). *J. Mater. Lett.*, **63**, 569.

17. Alipour, S.M., Nouri, M., Mokhtari, J., and Bahrami, S.H. (2009). *J. Carbohydr. Res.*, **344**, 2496.

18. Amiraliyan, N., Nouri, N., and Haghighatkish, M. (2009). *Fibers and Polymers*, **10**, 167.

19. Huang, Z.M., Zhang, Y.Z., Kotaki, M., and Ramakrishna, S. (2003). *J. Compos. Sci. Technol.*, **63**, 222.

# 7

1. Stevens, C. and Verhé, R. (2004). *Renewable bioresources: Scope and modification for non-food applications.* Wiley, London.

2. Parry, D.A.D. and Baker, E.N. (1984). Biopolymers. *Rep. Prog. Phys.*, **47**, 1133–1232.

3. Bhattacharyya, S., Guillot, S., Dabboue, H., Tranchant, J.-F., and Salvetat, J.-P. (2008). Carbon nanotubes as structural nanofibers for hyaluronic acid hydrogel scaffolds. *Biomacromolecules*, **9**, 505–509.

4. Marino Lavorgna, F.P., Mangiacapra, P., and Buonocore, G.G. (2010). Study of the combined effect of both clay and glycerol plasticizer on the properties of chitosan films. *Carbohydr. Polymer.*, **82**, 291–298.

5. Cao, X., Chen, Y., Chang, P.R., and Huneault, M.A. (2007). Preparation and properties of plasticized starch/multiwalled carbon nanotubes composites. *J. Appl. Polym. Sci.*, **106**, 1431–1437.

6. Ramakrishna, S., Mayer, J., Wintermantel, E., and Leong, K.W. (2001). Biomedical applications of polymer-composite materials: A review. *Comp. Sci. Technol.*, **61**, 1189–1224.

7. Liang, D., Hsiao, B.S., and Chu, B. (2007). Functional electrospun nanofibrous scaffolds for biomedical applications. *Adv. Drug Del. Rev.*, **59**, 1392–1412.

8. Liu, Z., Jiao, Y., Wang, Y., Zhou, C., and Zhang, Z. (2008). Polysaccharides-based nanoparticles as drug delivery systems. *Adv. Drug Del. Rev.*, **60**, 1650–1662.

9.  Agboh, O.C. and Qin, Y. (1997). Chitin and chitosan fibers. *Polym. Adv. Tech.*, **8**, 355–365.

10. Aranaz, I., Mengíbar, M., Harris, R., Paños, I., Miralles, B., Acosta, N., Galed, G., and Heras, Á. (2009). Functional characterization of chitin and chitosan. *Curr. Chem. Bio.*, **3**, 203–230.

11. Zhang, Y., Xue, C., Xue, Y., Gao, R., and Zhang, X. (2005). Determination of the degree of deacetylation of chitin and chitosan by X-ray powder diffraction. *Carbohy. Res.*, **340**, 1914–1917.

12. VandeVord, P.J., Matthew, H.W.T., DeSilva, S.P., Mayton, L., Wu, B., and Wooley, P.H. (2002). Evaluation of the biocompatibility of a chitosan scaffold in mice. *J. Biomed. Mater. Res.*, **59**, 585–590.

13. Ratajska, M., Strobin, G., Wiśniewska-Wrona, M., Ciechańska, D., Struszczyk, H., Boryniec, S., Biniaś, D., and Biniaś, W., (2003). Studies on the biodegradation of chitosan in an aqueous medium. *7 FIBRES & TEXT. East. Eur.*, **11**, 75–79.

14. Jayakumar, R., Prabaharan, M., Nair, S.V., and Tamura, H. (2010). Novel chitin and chitosan nanofibers in biomedical applications. *Biotechnol. Adv.*, **28**, 142–150.

15. Bamgbose, J.T., Adewuyi, S., Bamgbose, O., and Adetoye, A.A. (2010). Adsorption kinetics of cadmium and lead by chitosan. *Afr. J. Biotechnol.*, **9**, 2560–2565.

16. Krajewska, B. (2004). Application of chitin- and chitosan-based materials for enzyme immobilizations: A review. *Enzyme and Micro. Tech.*, **35**, 126–139.

17. Ueno, H., Mori, T., and Fujinaga, T. (2001). Topical formulations and wound healing applications of chitosan. *Adv. Drug Del. Rev.*, **52**, 105–115.

18. Kim, I.-Y., Seo, S.-J., Moon, H.-S., Yoo, M.-K., Park, I.-Y., Kim, B.-C., and Cho, C.-S. (2008). Chitosan and its derivatives for tissue engineering applications. *Biotechnol. Adv.*, **26**, 1–21.

19. Sinha, V.R., Singla, A.K., Wadhawan, S., Kaushik, R., Kumria, R., Bansal, K., and Dhawan, S. (2004). Chitosan microspheres as a potential carrier for drugs. *Int. J. Pharma.*, **274**, 1–33.

20. Martino, A.D., Sittinger, M., and Risbud, M.V. (2005). Chitosan: A versatile biopolymer for orthopaedic tissue-engineering. *Biomaterials*, **26**, 5983–5990.

21. Muzzarelli, R.A.A. (1997). Human enzymatic activities related to the therapeutic administration of chitin derivatives. *Cell. mol. life sci.*, **53**, 131–140.

22. Muzzarelli, R.A.A., Muzzarelli, C., Tarsi, R., Miliani, M., Gabbanelli, F., and Cartolari, M. (2001). Fungistatic activity of modified chitosans against saprolegnia parasitica. *Biomacromolecules*, **2**, 165–169.

23. Dutta, P.K., Tripathi, S., Mehrotra, G.K., and Dutta, J. (2009). Perspectives for chitosan based antimicrobial films in food applications. *Food Chem.*, **114**, 1173–1182.

24. Boonlertnirun, S., Boonraung, C., and Suvanasara, R. (2008). Application of chitosan in rice production. *J. Met. Mat. Min.*, **18**, 47–52.

25. Huang, K.-S., Wu, W.-J., Chen, J.-B., and Lian, H.-S. (2008). Application of low-molecular-weight chitosan in durable press finishing. *Carbohy. Polym.*, **73**, 254–260.

26. Lertsutthiwong, P., Chandrkrachang, S., and Stevens, W.F. (2000). The effect of the utilization of chitosan on properties of paper. *J. Met. Mat. Min.*, **10**, 43–52.

27. Subban, R.H.Y. and Arof, A.K. (1996). Sodium iodide added chitosan electrolyte film for polymer batteries. *Physica Scripta.*, **53**, 382–384.

28. Ottøy, M.H., Vårum, K.M., Christensen, B.E., Anthonsen, M.W., and Smidsrød, O. (1996). Preparative and analytical size-exclusion chromatography of chitosans. *Carbohy. Polym.*, **31**, 253–261.

29. Dutta, P.K., Dutta, J., and Tripathi, V.S. (2004). Chitin and chitosan: Chemistry, properties and applications. *JSIR*, **63**, 20–31.

30. Li, Q., Zhou, J., and Zhang, L. (2009). Structure and properties of the nanocomposite films of chitosan reinforced with cellulose whiskers. *J. Polymer Sci. B Polymer Phys.*, **47**, 1069–1077.

31. Thostenson, E.T., Li, C., and Chou, T.-W. (2005). Nanocomposites in context. *Comp. Sci. Technol.*, **65**, 491–516.

32. Iijima, S. (1991). Helical microtubules of graphitic carbon. *Nature*, **354**, 56–58.

33. Benthune, D.S., Kiang, C.H., Vries, M.S.d., Gorman, G., Savoy, R., Vazquez, J., and Beyers, R. (1993). Cobalt-catalysed growth of carbon nanotubes with single-atomic-layer-walls. *Nature*, **363**, 605–608.

34. Iijima, S. and Ichihashi, T. (1993). Single-shell carbon nanotubes of 1-nm diameter. *nature*, **363**, 603–605.

35. Trojanowicz, M. (2006). Analytical applications of carbon nanotubes: A review. *TrAC Trends Anal. Chem.*, **25**, 480–489.

36. Duclaux, L. (2002). Review of the doping of carbon nanotubes (multiwalled and single-walled). *Carbon*, **40**, 1751–1764.

37. Wang, Y.Y., Gupta, S., Garguilo, J.M., Liu, Z.J., Qin, L.C., and Nemanich, R.J. (2005). Growth and field emission properties of small diameter carbon nanotube films. *Diamond Relat. Mat.*, **14**, 714–718.

38. Guo, J., Datta, S., and Lundstrom, M. (2004). A numerical study of scaling issues for Schottky-Barrier carbon nanotube transistors. *IEEE Trans. Electron. Dev.*, **51**, 172–177.

39. Kuo, C.-S., Bai, A., Huang, C.-M., Li, Y.-Y., Hu, C.-C., and Chen, C.-C. (2005). Diameter control of multiwalled carbon nanotubes using experimental strategies. *Carbon*, **43**, 2760–2768.

40. Jacobsen, R.L., Tritt, T.M., Guth, J.R., Ehrlich, A.C., and Gillespie, D.J. (1995). Mechanical properties of vapor-grown carbon fiber. *Carbon*, **33**, 1217–1221.

41. Journet, C., Maser, W.K., Bernier, P., Loiseau, A., Chapelle, M.L.l., Lefrant, S., Deniard, P., Leek, R., and Fischer, J.E. (1997). Large-scale production of single-walled carbon nanotubes by the electric-arc technique. *Nature*, **388**, 756–758.

42. Thess, A., Lee, R., Nikolaev, P., Dai, H., Petit, P., Robert, J., Xu, C., Lee, Y.H., Kim, S.G., Rinzler, A.G., Colbert, D.T., Scuseria, G.E., Tomanek, D., Fischer, J.E., and Smalley, R.E. (1996). Crystalline ropes of metallic carbon nanotubes. *Science*, **273**, 483–487.

43. Cassell, A.M., Raymakers, J.A., Kong, J., and Dai, H. (1999). Large scale CVD synthesis of single-walled carbon nanotubes. *J. Phys. Chem. B*, **103**, 6482–6492.

44. Fan, S., Liang, W., Dang, H., Franklin, N., Tombler, T., Chapline, M., and Dai, H. (2000). Carbon nanotube arrays on silicon substrates and their possible application. *Physica E: Low-dimensional Sys. Nanostruc.*, **8**, 179–183.

45. Xie, S., Li, W., Pan, Z., Chang, B., and Sun, L. (2000). Carbon nanotube arrays. *Mat. Sci. Eng. A*, **286**, 11–15.

46. Tang, Z.K., Zhang, L., Wang, N., Zhang, X.X., Wen, G.H., Li, G.D., Wang, J.N., Chan, C.T., and Sheng, P. (2001). Superconductivity in 4 angstrom single-walled carbon nanotubes. *Science*, **292**, 2462–2465.

47. Peigney, A., Laurent, C., Flahaut, E., Bacsa, R.R., and Rousset, A. (2001). Specific surface area of carbon nanotubes and bundles of carbon nanotubes. *Carbon*, **39**, 507–514.

48. Pan, Z.W., Xie, S.S., Lu, L., Chang, B.H., Sun, L.F., Zhou, W.Y., Wang, G., and Zhang, D.L. (1999). Tensile tests of ropes of very long aligned multiwall carbon nanotubes. *Appl. Phys. Lett.*, **74**, 3152–3154.

49. Forro, L., Salvetat, J.P., Bonard, J.M., Basca, R., Thomson, N.H., Garaj, S., Thien-Nga, L., Gaal, R., Kulik, A., Ruzicka, B., Degiorgi, L., Bachtold, A., Schonenberger, C., Pekker, S., and Hernadi, K. (2000). Electronic and mechanical properties of carbon nanotubes. *Sci. and App. Nanot.*, 297–320.

50. Frank, S., Poncharal, P., Wang, Z.L., and Heer, W.A.d. (1998). Carbon nanotube quantum resistors. *Science*, **280**, 1744–1746.

51. Britzab, D.A. and Khlobystov, A.N. (2006). Noncovalent interactions of molecules with single walled carbon nanotubes. *Chem. Soc. Rev.*, **35**, 637–659.

52. Andrews, R. and Weisenberger, M.C. (2004). Carbon nanotube polymer composites. *Curr. Opin. Solid State Mater. Sci.*, **8**, 31–37.

53. Hirsch, A. (2002). Functionalization of single-walled carbon nanotubes. *Angew. Chem. Int. Ed.*, **41**, 1853–1859.

54. Firme III, C.P. and Bandaru, P.R. (2010). Toxicity issues in the application of carbon nanotubes to biological systems. *Nanomedicine: NBM*, **6**, 245–256.

55. Niyogi, S., Hamon, M.A., Hu, H., Zhao, B., Bhowmik, P., Sen, R., Itkis, M.E., and Haddon, R.C. (2002). Chemistry of single-walled carbon nanotubes. *Acc. Chem. Res.*, **35**, 1105–1113.

56. Kuzmany, H., Kukovecz, A., Simona, F., Holzweber, M., Kramberger, Ch., Pichler, T. (2004). Functionalization of carbon nanotubes. *Syn. Met.*, **141**, 113–122.

57. Spitalsky, Z., Tasis, D., Papagelis, K., and Galiotis, C. (2010). Carbon nanotube-polymer composites: Chemistry, processing, mechanical and electrical properties. *Prog. Polym. Sci.*, **35**, 357–401.

58. Narain, R., Housni, A., and Lane, L. (2006). Modification of carboxyl-functionalized single-walled carbon nanotubes with biocompatible, water-soluble phosphorylcholine and sugar-based polymers: Bioinspired nanorods. *J. Polym. Sci. A Polym. Chem.*, **44**, 6558–6568.

59. Wang, M., Pramoda, K.P., and Goh, S.H. (2006). Enhancement of interfacial adhesion and dynamic mechanical properties of poly(methyl methacrylate)/multiwalled carbon nanotube composites with amine-terminated poly(ethylene oxide). *Carbon*, **44**, 613–617.

60. Inahara, J., Touhara H, Mizuno T, et al., (2002). Fluorination of cup-stacked carbon nanotubes, structure and properties. *Fluorine Chem.*, **114**, 181–188.

61. Zhang, W., Sprafke, J.K., Ma, M., Tsui, E.Y., Sydlik, S.A., Rutledge, G.C., and Swager, T.M. (2009). Modular functionalization of carbon nanotubes and fullerenes. *J. Am. Chem. Soc.*, **131**, 8446–8454.

62. Peng He, Y.G., Lian, J., Wang, L., Qian, D., Zhao, J., Wang, W., Schulz, M.J., Zhou, X.P., and Shi, D. (2006). Surface modification and ultrasonication effect on the mechanical properties of carbon nanofiber/polycarbonate composites. *Comp. Part A: Appl. Sci. Manuf.*, **37**, 1270–1275.

63. Sulong, A.B., Azhari, C.H., Zulkifli, R., Othman, M.R., and Park, J. (2009). A comparison of defects produced on oxidation of carbon nanotubes by acid and UV ozone treatment. *Eur. J. Sci. Res.*, **33**, 295–304.

64. Wang, C., Guo, Z.-X., Fu, S., Wu, W., and Zhu, D. (2004). Polymers containing fuller-ene or carbon nanotube structures. *Prog. Polym. Sci.*, **29**, 1079–1141.

65. Rausch, J., Zhuang, R.-C., and Mäder, E. (2010). Surfactant assisted dispersion of functionalized multi-walled carbon nanotubes in aqueous media. *Comp. Part A: Appl. Sci. Manuf.*, **41**, 1038–1046.

66. Sahoo, N.G., Rana, S., Cho, J.W., Li, L., and Chan, S.H. (2010). Polymer nanocomposites based on functionalized carbon nanotubes. *Prog. Polym. Sci.*, **35**, 837–867.

67. Zheng, D., Li, X., and Ye, J. (2009). Adsorption and release behavior of bare and DNA-wrapped-carbon nanotubes on self-assembled monolayer surface. *Bioelectrochemistry*, **74**, 240–245.

68. Zhang, X., Meng, L., and Lu, Q. (2009). Cell behaviors on polysaccharide- wrapped single-wall carbon nanotubes: A quantitative study of the surface properties of biomimetic nanofibrous scaffolds. *ACS Nano*, **10**, 3200–3206.

69. Wang, H. (2009). Dispersing carbon nanotubes using surfactants. *Curr. Opin. Coll. Int. Sci.*, **14**, 364–371.

70. Liang, D., Hsiao, B.S., and Chu, B. (2007). Functional electrospun nanofibrous scaffolds for biomedical applications. *Adv. Drug Del. Rev.*, **59**, 1392–1412.

71. Ma, P.-C., Siddiqui, N.A., Marom, G., and Kim, J.-K. (2010). Dispersion and functionalization of carbon nanotubes for polymer-based nanocomposites: A review. *Comp. Part A: Appl. Sci. Manuf.*, **41**, 1345–1367.

72. Mottaghitalab, V., Spinks, G.M., and Wallace, G.G. (2005). The influence of carbon nanotubes on mechanical and electrical properties of polyaniline fibers. *Syn. Met.*, **152**, 77–80.

73. Cheung, W., Pontoriero, F., Taratula, O., Chen, A.M., and He, H. (2010). DNA and carbon nanotubes as medicine. *Adv. Drug Del. Rev.*, **62**, 633–649.

74. Piovesan, S., Cox, P.A., Smith, J.R., Fatouros, D.G., and Roldo, M. (2010). Novel biocompatible chitosan decorated single-walled carbon nanotubes (SWNTs) for biomedical applications: Theoretical and experimental investigations. *Phys. Chem. Chem. Phys.*, **12**, 15636–15643.

75. Chen, J., Liu, H., Weimer, W.A., Halls, M.D., Waldeck, D.H., and Walker, G.C. (2002). Noncovalent engineering of carbon nanotube surfaces by rigid, functional conjugated polymers. *J. Am. Chem. Soc.*, **124**, 9034–9035.

76. Vamvakaki, V., Fouskaki, M., and Chaniotakis, N. (2007). Electrochemical biosensing systems based on carbon nanotubes and carbon nanofibers. *Anal. Lett.*, **40**, 2271–2287.

77. Kang, Y., Liu, Y.-C., Wang, Q., Shen, J.-W., Wu, T., and Guan, W.-J. (2009). On the spontaneous encapsulation of proteins in carbon nanotubes. *Biomaterials*, **30**, 2807–2815.

78. Harrison, B.S. and Atala, A. (2007). Carbon nanotube applications for tissue engineering. *Biomaterials*, **28**, 344–353.

79. Moniruzzaman, M. and Winey, K.I. (2006). Polymer nanocomposites containing carbon nanotubes. *Macromolecules*, **39**, 5194–5205.

80. Ajayan, P.M., Stephan, O., Colliex, C., and Trauth, D. (1994). Aligned carbon nanotube arrays formed by cutting a polymer resin-nanotube composite. *Science*, **265**, 1212–1214.

81. Liu, P. (2005). Modifications of carbon nanotubes with polymers. *Eur. Polym. J.*, **41**, 2693–2703.

82. Manchado, M.A.L., Valentini, L., Biagiotti, J., and Kenny, J.M. (2005). Thermal and mechanical properties of single-walled carbon nanotubes-polypropylene composites prepared by melt processing. *Carbon*, **43**, 1499–1505.

83. Wang, Q. and Varadan, V.K. (2005). Stability analysis of carbon nanotubes via continuum models. *Smart Mater. Struct.*, **14**, 281–286.

84. E. Materials Information. (2004). *Fiber reinforced composites*. CSA Journal Division.

85. Advani, S.G. (2006). *Processing and properties of nanocomposites*. World Scientific Publishing Company.

86. Nan, C.W., Shi, Z., and Lin, Y. (2003). A simple model for thermal conductivity of carbon nanotube-based composites. *Chem. Phys. Lett.*, **375**, 666–669.

87. Coleman, J.N., Khan, U., Blau, W.J., and Gun'ko, Y.K. (2006). Small but strong: A review of the mechanical properties of carbon nanotube-polymer composites. *Carbon*, **44**, 1624–1652.

88. Wagner, H.D., Lourie, O., Feldman, Y., and Tenne, R. (1998). Stress-induced fragmentation of multiwall carbon nanotubes in a polymer matrix. *Appl. Phys. Lett.*, **72**, 188–190.

89. Wang, Q., Dai, J., Li, W., Wei, Z., and Jiang, J. (2008). The effects of CNT alignment on electrical conductivity and mechanical properties of SWNT/epoxy nanocomposites. *Comp. Sci. Technol.*, **68**, 1644–1648.

90. Khan, U., Ryan, K., Blau, W.J., and Coleman, J.N. (2007). The effect of solvent choice on the mechanical properties of carbon nanotube-polymer composites. *Comp. Sci. Technol.*, **67**, 3158–3167.

91. Allaoui, A., Bai, S., Cheng, H.M., and Bai, J.B. (2002). Mechanical and electrical properties of a MWNT/epoxy composite. *Comp. Sci. Technol.*, **62**, 1993–1998.

92. Esawi, A.M.K. and Farag, M.M. (2007). Carbon nanotube reinforced composites: Potential and current challenges. *Materials & Design*, **28**, 2394–2401.

93. Kamaras, K., Itkis, M.E., Hu, H., Zhao, B., and Haddon, R.C. (2003). Covalent bond formation to a carbon nanotube metal. *Science*, **301**, 1501.

94. Shenogin, S., Bodapati, A., Xue, L., Ozisik, R., and Keblinski, P. (2004). Effect of chemical functionalization on thermal transport of carbon nanotube composites. *Appl. Phys. Lett.*, **85**, 2229–2231.

95. MacDonald, R.A., Laurenzi, B.F., Viswanathan, G., Ajayan, P.M., and Stegemann, J.P. (2005). Collagen–carbon nanotube composite materials as scaffolds in tissue engineering. *J. Biomed. Mater. Res. Part A*, **74A**, 489–496.

96. Suryasarathi Bose, R.A.K. and Moldenaers, P. (2010). Assessing the strengths and weaknesses of various types of pre-treatments of carbon nanotubes on the properties of polymer/carbon nanotubes composites: A critical review. *Polymer*, **51**, 975–993.

97. Gojny, F.H., Wichmann, M.H.G., Fiedler, B., Kinloch, I.A., Bauhofer, W., Windle, A.H., and Schulte, K. (2006). Evaluation and identification of electrical and thermal

conduction mechanisms in carbon nanotube/epoxy composites. *Polymer*, **47**, 2036–2045.

98. Tosun, Z. and McFetridge, P.S. (2010). A composite SWNT–collagen matrix: Characterization and preliminary assessment as a conductive peripheral nerve regeneration matrix. *J. Neural. Eng.*, **7**, 10.

99. Wang, S.-F., Shen, L., Zhang, W.-D., and Tong, Y.-J. (2005). Preparation and mechanical properties of chitosan/carbon nanotubes composites. *Biomacromolecules*, **6**, 3067–3072.

100. Liu, Y.-L., Chen, W.-H., and Chang, Y.-H. (2009). Preparation and properties of chitosan/carbon nanotube nanocomposites using poly (styrene sulfonic acid)-modified CNTs. *Carbohy. Polym.*, **76**, 232–238.

101. Spinks, G.M., Shin, S.R., Wallace, G.G., Whitten, P.G., Kim, S.I., and Kim, S.J. (2006). Mechanical properties of chitosan/CNT microfibers obtained with improved dispersion. *Sens. Actuators B: Chem.*, **115**, 678–684.

102. Gandhi, M., Yang, H., Shor, L., and Ko, F. (2009). Post-spinning modification of electrospun nanofiber nanocomposite from Bombyx mori silk and carbon nanotubes. *Polymer*, **50**, 1918–1924.

103. Rahatekar, S.S., Rasheed, A., Jain, R., Zammarano, M., Koziol, K.K., Windle, A.H., Gilman, J.W., and Kumar, S. (2009). Solution spinning of cellulose carbon nanotube composites using room temperature ionic liquids. *Polymer*, **50**, 4577–4583.

104. Zhang, H., Wang, Z., Zhang, Z., Wu, J., Zhang, J., and He, J. (2007). Regenerated-cellulose/multiwalled-carbon-nanotube composite fibers with enhanced mechanical properties prepared with the ionic liquid 1-allyl-3-methylimidazolium chloride. *Adv. Mater.*, **19**, 698–704.

105. Lu, P. and Hsieh, Y.-L. (2010). Multiwalled carbon nanotube (MWCNT) reinforced cellulose fibers by electrospinning. *Appl. Mater. & Interf.*, **2**, 2413–2420.

106. Wang, Y. and Yeow, J.T.W. (2009). A review of carbon nanotubes-based gas sensors. *J. Sens.*,1–24.

107. Yang, W., Thordarson,P., Gooding, J.J., Ringer, S.P., and Braet, F. (2007). Carbon nanotubes for biological and biomedical applications. *Nanotechnology*, **18**, 12.

108. Wang, J. (2005). Carbon-nanotube based electrochemical biosensors: A review. *Electroanalysis*, **17**, 7–14.

109. Foldvari, M. and Bagonluri, M. (2008). Carbon nanotubes as functional excipients for nanomedicines: II. Drug delivery and biocompatibility issues. *Nanomedicine: NBM*, **4**, 183–200.

110. Belluccia, S., Balasubramanianab, C., Micciullaac, F., and Rinaldid, G. (2007). CNT composites for aerospace applications. *J. Exp. Nanosci.*, **2**, 193–206.

111. Kang, I., Heung, Y.Y., Kim, J.H., Lee, J.W., Gollapudi, R., Subramaniam, S., Narasimhadevara, S., Hurd, D., Kirikera, G.R., Shanov, V., Schulz, M.J., Shi, D., Boerio, J., Mall, S., and Ruggles-Wren, M. (2006). Introduction to carbon nanotube and nanofiber smart materials. *Comp. Part B*, **37**, 382–394.

112. Mottaghitalab, V., Spinks, G.M., and Wallace, G.G. (2006). The development and characterisation of polyaniline--single walled carbon nanotube composite fibres using 2-acrylamido-2 methyl-1-propane sulfonic acid (AMPSA) through one step wet spinning process. *Polymer*, **47**, 4996–5002.

113. Wang, C.Y., Mottaghitalab, V., Too, C.O., Spinks, G.M., and Wallace, G.G. (2007). Polyaniline and polyaniline-carbon nanotube composite fibres as battery materials in ionic liquid electrolyte. *J. Power Sour.*, **163**, 1105–1109.

114. Mottaghitalab, V., Xi, B., Spinks, G.M., and Wallace, G.G. (2006). Polyaniline fibres containing single walled carbon nanotubes: Enhanced performance artificial muscles. *Syn. Met.*, **156**, 796–803.

115. Ong, Y.T., Ahmad, A.L., Zein, S.H.S., and Tan, S.H. (2010). A review on carbon nanotubes in an enviromental protection and green engineering perspective. *Braz. J. Chem. Eng.*, **27**, 227–242.

116. Sariciftci, N.S., Smilowitz, L., Heeger, A.J., and Wudi, F. (1992). Photoinduced electron transfer from a conducting polymer to buckminsterfullerene. *Science*, **258**, 1474–1478.

117. Harris, P.J.F. (2004). Carbon nanotube composites. *Inter. Mater. Rev.*, **49**, 31–43.

118. Azeredo, H.M.C.d. (2009). Nanocomposites for food packaging applications. *Food Res. Int.*, **42**, 1240–1253.

119. Phuoc, T.X., Massoudi, M., and Chen, R.-H. (2011). Viscosity and thermal conductivity of nanofluids containing multi-walled carbon nanotubes stabilized by chitosan. *Int. J. Therm. Sci.*, **50**, 12–18.

120. Zhang, M., Smith, A., and Gorski, W. (2004). Carbon nanotube-chitosan system for electrochemical sensing based on dehydrogenase enzymes. *Anal. Chem.*, **76**, 5045–5050.

121. Liu, Y., Wang, M., Zhao, F., Xu, Z., and Dong, S. (2005). The direct electron transfer of glucose oxidase and glucose biosensor based on carbon nanotubes/chitosan matrix. *Biosensors and Bioelectronics*, **21**, 984–988.

122. Tkac, J., Whittaker, J.W., and Ruzgas, T. (2007). The use of single walled carbon nanotubes dispersed in a chitosan matrix for preparation of a galactose biosensor. *Biosensors and Bioelectronics*, **22**, 1820–1824.

123. Tsai, Y.-C., Chen, S.-Y., and Liaw, H.-W. (2007). Immobilization of lactate dehydrogenase within multiwalled carbon nanotube-chitosan nanocomposite for application to lactate biosensors. *Sens Actuators B: Chem.*, **125**, 474–481.

124. Zhou, Y., Yang, H., and Chen, H.-Y. (2008). Direct electrochemistry and reagentless biosensing of glucose oxidase immobilized on chitosan wrapped single-walled carbon nanotubes. *Talanta*, **76**, 419–423.

125. Li, J., Liu, Q., Liu, Y., Liu, S., and Yao, S. (2005). DNA biosensor based on chitosan film doped with carbon nanotubes. *Anal. Biochem.*, **346**, 107–114.

126. Bollo, S., Ferreyr, N.F., and Rivasb, G.A. (2007). Electrooxidation of DNA at glassy carbon electrodes modified with multiwall carbon nanotubes dispersed in chitosan. *Electroanalysis*, **19**, 833–840.

127. Zeng, Y., Zhu, Z.-H., Wang, R.-X., and Lu, G.-H. (2005). Electrochemical determination of bromide at a multiwall carbon nanotubes-chitosan modified electrode. *Electrochim. Acta*, **51**, 649–654.

128. Qian, L. and Yang, X. (2006). Composite film of carbon nanotubes and chitosan for preparation of amperometric hydrogen peroxide biosensor. *Talanta*, **68**, 721–727.

129. Liu, Y., Qu, X., Guo, H., Chen, H., Liu, B., and Dong, S. (2006). Facile preparation of amperometric laccase biosensor with multifunction based on the matrix of carbon nanotubes-chitosan composite. *Biosensors and Bioelectronics*, **21**, 2195–2201.

130. Naficy, S., Razal, J.M., Spinks, G.M., and Wallace, G.G. (2009). Modulated release of dexamethasone from chitosan-carbon nanotube films. *Sens. Actuators A: Phys.*, **155**, 120–124.

131. Yang, J., Yao, Z., Tang, C., Darvell, B.W., Zhang, H., Pan, L., Liu, J., and Chen, Z. (2009). Growth of apatite on chitosan-multiwall carbon nanotube composite membranes. *Appl. Surf. Sci.*, **255**, 8551–8555.

132. Kaushik, A., Solanki, P.R., Pandey, M.K., Kaneto, K., Ahmad, S., and Malhotra, B.D. (2010). Carbon nanotubes—chitosan nanobiocomposite for immunosensor. *Thin Solid Films*, **519**, 1160–1166.

133. Yang, H., Yuan, R., Chai, Y., and Zhuo, Y. (2011). Electrochemically deposited nanocomposite of chitosan and carbon nanotubes for detection of human chorionic gonadotrophin. *Coll. Surf. B: Bioint.*, **82**, 463–469.

134. Zheng, Y.Q.C.W. and Zheng, Y.F. (2008). Adsorption and electrochemistry of hemoglobin on chi-carbon nanotubes composite film. *Appl. Surf. Sci.*, **255**, 571–573.

135. Tang, C., Zhang, Q., Wang, K., Fu, Q., and Zhang, C. (2009). Water transport behavior of chitosan porous membranes containing multi-walled carbon nanotubes (MWNTs). *J. Membrane Sci.*, **337**, 240–247.

136. Wei-Hong, C., Ying-Ling, L., and Yu-Hsun, C. (2009). Preparation and properties of chitosan/carbon nanotube nanocomposites using poly(styrene sulfonic acid)-modified CNTs. *Carbohy. Polym.*, **76**, 232–238.

137. Ghica, M.E., Pauliukaite, R., Fatibello-Filho, O., and Brett, C.M.A. (2009). Application of functionalised carbon nanotubes immobilised into chitosan films in amperometric enzyme biosensors. *Sens. Actuators B: Chem.*, **142**, 308–315.

138. Kandimalla, V.B. and Ju, H. (2006). Binding of acetylcholinesterase to multiwall carbon nanotube-cross- linked chitosan composite for flow-injection amperometric detection of an organophosphorous insecticide. *Chem. Eur. J.*, **12**, 1074–1080.

139. Du, D., Huang, X., Cai, J., Zhang, A., Ding, J., and Chen, S. (2007). An amperometric acetylthiocholine sensor based on immobilization of acetylcholinesterase on a multiwall carbon nanotube–cross-linked chitosan composite. *Anal. Bioanal. Chem.*, **387**, 1059–1065.

140. Du, D., Huang, X., Cai, J., and Zhang, A. (2007). Amperometric detection of triazophos pesticide using acetylcholinesterase biosensor based on multiwall carbon nanotube-chitosan matrix. *Sens. Actuators B: Chem.*, **127**, 531–535.

141. Salam, M.A., Makki, M.S.I., and Abdelaal, M.Y.A. (2010). Preparation and characterization of multi-walled carbon nanotubes/chitosan nanocomposite and its application for the removal of heavy metals from aqueous solution. *J. Alloys Comp.*, **509**, 2582–2587.

142. Tang, J., Liu, Y., Chen, X., and Xin, J.H. (2005). Decoration of carbon nanotubes with chitosan. *Carbon*, **43**, 3178–3180.

143. Luo, X.-L., Xu, J.-J., Wang, J.-L., and Chen, H.-Y. (2005). Electrochemically deposited nanocomposite of chitosan and carbon nanotubes for biosensor application. *Chem. Comm.*, 2169–2171, DOI: 10.1039/b419197h

144. Tan, Y., Bin G., Xie, Q., Ma M., and Yao, S. (2009). Preparation of chitosan-dopamine-multiwalled carbon nanotubes nanocomposite for electrocatalytic oxidation and sensitive electroanalysis of NADH. *Sens. Actuators B: Chem.*, **137**, 547–554.

145. Shieh, Y.-T. and. Yang, Y.-F. (2006). Significant improvements in mechanical property and water stability of chitosan by carbon nanotubes. *Eur. Polym. J.*, **42**, 3162–3170.

146. Wu, Z., Feng, W., Feng, Y., Liu, Q., Xu, X., Sekino, T., Fujii, A., and Ozaki, M. (2007). Preparation and characterization of chitosan-grafted multiwalled carbon nanotubes and their electrochemical properties. *Carbon*, **45**, 1212–1218.

147. Carson, L., Kelly-Brown, C., Stewart, M., Oki, A., Regisford, G., Luo, Z., and Bakhmutov, V.I. (2009). Synthesis and characterization of chitosan-carbon nanotube composites. *Mater. Lett.*, **63**, 617–620.

148. Ke, G., Guan, W.C., Tang, C.Y., Hu, Z., Guan, W.J., Zeng, D.L., and Deng, F. (2007). Covalent modification of multiwalled carbon nanotubes with a low molecular weight chitosan. *Chin. Chem. Lett.*, **18**, 361–364.

149. Baek, S.-H., Kim, B., and Suh, K.-D. (2008). Chitosan particle/multiwall carbon nanotube composites by electrostatic interactions. *Coll. Surf. A: Physicochem. Eng. Asp.*, **316**, 292–296.

150. Zhao, Q., Yin, J., Feng, X., Shi, Z., Ge, Z., and Jin, Z. (2010). A biocompatible chitosan composite containing phosphotungstic acid modified single-walled carbon nanotubes. *J. Nanosci. Nanotechnol.*, **10**, 1–4.

151. Yu, J.-G., Huang, K.-L., Tang, J.-C., Yang, Q., and Huang, D.-S. (2009). Rapid microwave synthesis of chitosan modified carbon nanotube composites. *Int. J. Bio. Macromol.*, **44**, 316–319.

152. Wang, Z.-k., Hu, Q.-l., and Cai, L. (2010). Chitosan and multiwalled carbon nanotube composite rods. *Chin. J. Polym. Sci.*, **28**, 801–806.

153. Xiao-ying, J. and Xiao-bo, L. (2010). Electrostatic layer-by-layer assembled multilayer films of chitosan and carbon nanotubes. *New Carbon Mater.*, **25**, 237–240.

154. Lau, C. and Cooney, M.J. (2008). Conductive macroporous composite chitosan-carbon nanotube scaffolds. *Langmuir*, **24**, 7004–7010.

155. Zhang, W., Sprafke, J.K., Ma, M., Tsui, E.Y., Sydlik, S.A., Rutledge, G.C., and Swager, T.M. (2009). Modular functionalization of carbon nanotubes and fullerenes. *J. Am. Chem. Soc.*, **131**, 8446–8454.

156. Razal, J.M., Gilmore, K.J., and Wallace, G.G. (2008). Carbon nanotube biofiber formation in a polymer-free coagulation bath. *Adv. Funct. Mater.*, **18**, 61–66.

157. Lynam, C., Moulton, S.E., and Wallace, G.G. (2007). Carbon-nanotube biofibers. *Adv. Mat.*, **19**, 1244–1248.

158. Spinks, G.M., Shin, S.R., Wallace, G.G., Whitten, P.G., Kim, S.I., and Kim, S.J.

(2006). Mechanical properties of chitosan/ CNT microfibers obtained with improved dispersion. *Sens. Actuators B: Chem.*, **115**, 678–684.

# 8

1. Wallace, G.G., Spinks, G.M., Kane-maguire, L.A.P., and Teasdale, P.R. (2003). Conductive electroactive polymers, 2nd ed., CRC, USA.

2. Kang, T.S., Lee, S.W., Joo, J., Lee, J.Y. (2005). *Synthetic Met.*, **153**, 61–64.

3. Malinauskas, A. (2000). *Polymer*, **42**,3957–3972.

4. Rossi, D.D., Sanata, A.D., and Mazzoldi, A. (1999). *Mater. Sci. Eng.*, **C7**, 31.

5. Huang, Z.M., Zhang, Y.Z., Kotaki, M., and Ramakrishna, S. (2003). *Compos. Sci. Tech.*, **63**, 2223–2253.

# 9

1. Hückel, E. (1930). Zur quantentheorie der doppelbindung [Quantum theory of double linkings]. *Z. Physik.*, **60**, 423–432.

2. Purrello, R., Gurrieri, S., and Lauceri, R. (1999). Porphyrin assemblies as chemical sensors. *Coord. Chem. Rev.*, **190**, 683–706.

3. Smith, K. (1975). *Porphyrins and metalloporphyrins*, Vol. I-VII, Elsevier, Amsterdam.

4. Sanders, J.K.M., Atwood, J.L., Davies, J.E.D., MacNicol, D.D., and Vogel, F. (1996). *Comprehensive supramolecular chemistry*, Vol. 9, Elsevier, Amsterdam.

5. Battersby, A.R., Fookes, O.J.R., Matcham, G.W.J., and McDonald, F. (1980). Biosynthesis of the pigments of life: Formation of the macrocycle. *Nature*, **285**, 17–21.

6. Kral, V., Kralova, J., Kaplanek, R., Briza, T., and Martasek, P. (2006). Quo vadis porphyrin chemistry? *Physiol. Res.*, **55**, S3–S26.

7. Wai-Yin Sun, R. and Che, C-M. (2009) The anti-cancer properties of gold (III) compounds with dianionic porphyrin and tetradentate ligands. *Coord. Chem.Rev.*, **253**, 1682–1691.

8. Huheey, J.E., Keiter, E.A., and Keiter, R.L. (2006). *Inorganic chemistry: Principles of structure and reactivity*, 4th ed., Pearson Education.

9. Harper, H.A., Rodwell, V.W., and Mayes, P.A. (1979). *Review of physiological chemistry*, 17th ed., Lange Medical Publications, Los Altos, CA, 235–237.

10. Rothmund, P. (1936). A new porphyrin synthesis. The synthesis of porphyrin. *J. Am. Chem. Soc.*, **58**, 625–627.

11. Zerbetto, F., Zgierski, M.Z., and Orlandi, G. (1987). Normal modes and ground state geometry of porphine. Evidence for dynamic instability of the D2h configuration. *Chem. Phys. Lett.*, **139**, 401–406.

12. Dolphin, D. (1979). *The porphyrins*, Vol. 1 & 2, Academic Press, New York.

13. Ghosh, A. (1998). First-principles quantum chemical studies of porphyrins. *Acc. Chem. Res.*, **31**, 189–198.

14. Mendez, F. and Gazquez, J.L. (1994). Chemical reactivity of enolate ions: The local hard and soft acids and bases principle viewpoint. *J. Am. Chem. Soc.*, **116**, 9298–9301.

15. Martin, R.B. (1998). Metal ion stabilities correlate with electron affinity rather than hardness or softness, *Inorg. Chem. Acta*, **283**, 30–56.

16. Ghosh, A. and Almlof, J. (1995). Structure and stability of cis-porphyrin. *J. Phys. Chem.*, **99**, 1073–1075.

17. Punnagai, M., Joseph, S., and Sastry, G.N. (2004). A theoretical study of porphyrin isomers and their core-modified analogues: *cis-trans* isomerism, tautomerism and relative stabilities. *J. Chem. Sci.*, **116**, 271–283.

18. Feng Xin-Tian, Yu Jian-Guo, Lei M., Fang Wei-Hai & Liu S. (2009). Toward understanding metal-binding specificity of porphyrin: A conceptual density functional theory study. *J. Phys. Chem. B*, **113**, 13381–13389.

19. Dewar, M.J.S., Zoebisch, E.G., Healy, E.F., and Stewart, J.J.P. (1985). Development and use of quantum mechanical molecular models. 76. AM1: A new general purpose quantum mechanical molecular model. *J. Am. Chem. Soc.*, **107**, 3902–3909.

20. Stewart, J.J.P. (1989). Optimization of parameters for semiempirical methods I. Method. *J. Comput. Chem.*, **10**, 209–220.

21. Hasanein, A.A. and Evans, M.W. (1996*). Computational methods in quantum chemistry*, Quantum Chemistry, Vol. 2, World Scientific Publishing, Singapore.

22. Parr, R.G., Donnely, A., Levy, M., and Palke, W. (1978). Electronegativity, The density functional viewpoint. *J. Chem. Phys.*, **68**, 3801–3807.

23. Gyftpoulous, E.P. and Hatsopoulos, G.N. (1968). Quantum thermodynamic definition of electronegativity. *Proc. Natl. Acad. Sci.*, **60**, 786–793.

24. Iczkowski, R.P. and Margrave, J.L. (1961). Electronegativity. *J. Am. Chem. Soc.*, **83**, 3547–3551.

25. Parr, R.G. and Pearson, R.G. (1983). Absolute hardness, companion parameter to absolute electronegativity. *J. Am. Chem. Soc.*, **105**, 7512–7516.

26. Pearson, R.G. (1986). Absolute electronegativity and hardness correlated with molecular orbital theory. *Proc. Natl. Acad. Sci.*, **83**, 8440–8441.

27. Parr, R.G. and Yang, W. (1984). Density functional approach to the frontier-electron theory of chemical reactivity. *J. Am. Chem. Soc.*, **106**, 4049–4050.

28. Maynard, A.T. and Covell, D.G. (2001). Reactivity of zinc finger cores: Analysis of protein packing and electrostatic screening. *J. Am. Chem. Soc.*, **123**, 1047–1058.

29. Geerlings, P., Proft, F.D., and Langenaeker, W. (2003). Conceptual density functional theory. *Chem. Rev.*, **103**, 1793–1874.

30. Yang, W. and Parr, R.G. (1985). Hardness, softness, and the Fukui function in the electronic theory of metals and catalysis. *Proc. Natl. Acad. Sci. USA*, **82**, 6723–6726.

31. Fukui, K., Yonezawa, T., and Shingu, H. (1952). A molecular orbital theory of reactivity in aromatic hydrocarbons. *J. Chem. Phys.*, **20**, 722–725.

32. Yang, W. and Mortier, W.J. (1986). The use of global and local molecular parameters for the analysis of the gas-phase basicity of amines. *J. Am. Chem. Soc.*, **108**, 5708–5711.

33. Li, Y. and Evans, J.N.S. (1995). The Fukui function: A key concept linking frontier molecular orbital theory and the hard-soft-acid-base principle. *J. Am. Chem. Soc.*, **117**, 7756–7759.

34. Chattaraj, P.K., Maity, B., and Sarkar, U. (2003). Philicity: A unified treatment of chemical reactivity and selectivity. *J. Phys. Chem. A*, **107**, 4973–4975.

35. Mulliken, R.S. (1955). Electronic population analysis on LCAO-MO molecular wave functions. I. *J. Chem. Phys.*, **23**, 1833–1840.

36. Csizmadia, G. (1976). *Theory and practice of MO calculations on organic molecules.* Elsevier, Amsterdam.

37. Pariser, R. and Parr, R.G. (1953). A semiempirical theory of the electronic spectra and electronic structure of complex unsaturated molecules. *J. Chem. Phys.*, **21**, 466–477.

38. Pople, J.A. (1953). Electron interaction in unsaturated hydrocarbons. *Trans. Faraday Soc.*, **49**, 1375–1384.

39. Parr, R.G. (1963). *Quantum theory of molecular electronic structure.* W.A. Benjamin, Inc., New York.

40. Pople, J.A. (1962). Reply to Letter by H.F. Hameka. *J. Chem. Phys.*, **37**, 3009.

41. Fischer-Hjalmars, I. (1965). Deduction of the zero differential overlap approximation from an orthogonal atomic orbital basis. *J. Chem. Phys.*, **42**, 1962–1972.

42. ArgusLab 4.0, M.A. Thompson, planaria software LLC, seattle, WA. http://www.arguslab.com

43. Torrent-Sucarrat, M., Proft, F.D., Geerlings, P., and Ayers, P.W. (2008). Do the local softness and hardness indicate the softest and hardest regions of a molecule? *Chem. Eur. J.*, **14**, 8652–8660.

44. Damoun, S., Van de woude, G., Mandez, F., and Geerlings, P. (1997). Local softness as a regioselectivity indicator in [4+2] cycloaddition reactions. *J. Phys. Chem. A*, **101**, 886–893.

45. Pearson, R.G. (1963). Hard and soft acids and bases. *J. Am. Chem. Soc.*, **85**, 3533–3539.

# Index